Color Film for Color Television

Image and Sound Technology

BASIC MOTION PICTURE TECHNOLOGY
L. Bernard Happé

WIDE-SCREEN CINEMA AND STEREOSCOPIC SOUND
Michael Z. Wysotsky

ACOUSTICS OF STUDIOS AND AUDITORIA
V. S. Mankovsky

COLOR FILM FOR COLOR TELEVISION

Rodger J. Ross

COMMUNICATION ARTS BOOKS
Hastings House, Publishers
New York N.Y. 10016

Library of Congress Catalog Card No. 79–110805

SBN 8038–1137–3

*Printed and bound in Great Britain at
the Pitman Press, Bath*

Contents

Introduction

Film has always been an important factor in television broadcasting, not only as a source of ready-made programs, but also as an original program recording medium.

Feature films made originally for showing in theaters and adapted for television use provide one of the most popular forms of broadcast entertainment. These film programs, together with great numbers of dramatic and comedy serials filmed especially for television, account today for 80 to 90 per cent of prime time on the major networks in the United States.

A 16 mm film camera is a wonderfully versatile device—light, easily portable, capable of being carried anywhere a man (or woman) can go. A roll of film stock loaded into the camera turns it into a combined camera-recorder. With the aid of this simple, inexpensive, trouble-free equipment, the very finest color recordings can be made for local programming purposes.

The rapid transition from monochrome to color broadcasting within the past few years was made possible to a large extent by the availability of large numbers of color films suitable for use in television programming. Many of the feature films produced in the past 15 or 20 years have been in color, and prior to the change-over to color by the television industry, these films were being transmitted to the public in monochrome. With so much program material available on color film, very modest capital investments in color film reproducing equipment enabled television stations to begin broadcasting in color.

Because for many years the motion picture industry has been working mainly in color, television programs that were being filmed with black-and-white negative materials before the change-over, could be—and were—

quickly converted to color simply by changing the film stock used in the cameras.

It is not at all difficult to demonstrate that film is capable of yielding excellent results in television reproduction when the picture and sound records have been made properly for this purpose. Many of the problems that broadcasters encounter in the use of film are due to misunderstanding or lack of appreciation of the role of film in television programming. Conventional motion picture production methods are too slow and cumbersome for television. Besides, motion picture making is a craft, and it is very difficult to maintain adequate uniformity of film image characteristics to satisfy the requirements of electronic reproduction.

A subject of perennial controversy in television circles is the question of 35 versus 16 mm film. In the early stages of television development in North America, broadcasters adopted the 16 mm size because it was cheaper and easier to handle than the professional 35 mm film used in theaters. The larger network centers installed 35 mm telecines so that they could accept 35 mm prints directly from motion picture laboratories. Individual television stations equipped only with 16 mm telecines require reduction prints made from the 35 mm originals. Alternatively, a 16 mm process may be used throughout, beginning with the original camera film, but professional motion picture producers and laboratories are very reluctant to become involved in the use of the smaller film format. A carefully made 16 mm color film should be perfectly acceptable for television reproduction, but it is certainly much easier to achieve the desired results with the larger film size.

In Europe, the use of 35 mm has been established as the primary film format, only partially supplemented by 16 mm. There are some indications of a trend towards more extensive use of the smaller film, a trend strongly influenced by economic considerations, as well as the significant improvements that are being achieved in 16 mm picture and sound quality.

Color offers new and exciting possibilities in television program production. By far the easiest and least expensive way for a television station to get into local color production is to make use of a 16 mm film camera loaded with color film. By applying proper control of exposure, processing, and reproduction in telecine, it should be possible to achieve a standard of picture quality at least as good as any other source of color programs. The simplest, easiest and best way of recording the accompanying sound is to make use of synchronously running perforated magnetic film, or film coated with a magnetic stripe in the sound track area.

The motion picture industry has developed for its own use an extensive array of color film materials and processes, capable of yielding the very finest color picture quality on large projection screens. All of these materials and processes are readily available for television use as well. However, equipment designed for motion picture production is, in many respects, not well suited for the production of television programs. Currently there is considerable interest in making use of electronic aids in the operation of film cameras.

To undertake multiple camera operation, a producer* requires monitoring facilities providing greatly enlarged displays of the images appearing in the camera viewfinders. Remote control of focus is essential; at the same time, proper correlation must be established between the level of exposure of the film in the camera and the amplitude of the video signals being monitored at the camera control position.

To make the most effective use of film in television, it is essential to adopt the concept of a standard system for the recording and reproduction of pictures and sound. First of all, a standard telecine setting-up procedure is required. This will establish the basic film image requirements for television reproduction. The next step is to organize a standard film recording procedure by which images with the desired characteristics can be obtained consistently. This should not be difficult with currently available color film materials and processing facilities.

This is a time of rapid change. New ideas are being introduced, and equipment is being developed that quickly outdates yesterday's well-established and familiar practices. Only a few years ago 16 mm was considered to be sub-standard, unsuited for serious professional use. Today, broadcasters are intrigued with the possibilities of automated television station operation utilizing 8 mm film cassettes. The future of film in television is very bright indeed, but before the full potential of this medium can be realized, there needs to be a better understanding on the part of film makers of television's technical requirements. At the same time, there needs to be a better appreciation of film's advantages on the part of television broadcasters.

The purpose of this book is to show, in a practical way, how easy it is to produce high quality color television programs with color film. Currently available color films and processes are described in sufficient detail to enable television producers to select suitable materials for different types of program requirements. For the most part, the materials and methods outlined in this book are already in everyday use; at the same time, advanced methods of program production on film, as yet largely experimental and in some cases still untried, have been suggested.

* The term "producer" is used throughout this book in referring to the individual responsible for program production. The producer, in television as well as motion picture operations, has overall responsibility for a production, while a director is in charge of people and technical facilities. When these functions are combined the term "producer-director" is often used.

1 Color fundamentals

Color is a visual phenomenon, created in the eye by the action of radiant energy. Different color sensations are caused by different kinds of energy. This energy comes from the sun; from metals heated to incandescence, such as tungsten lamps; from electric arcs, and chemical substances such as phosphors. The small portion of the electromagnetic spectrum to which the eye is sensitive is shown in Fig. 1.

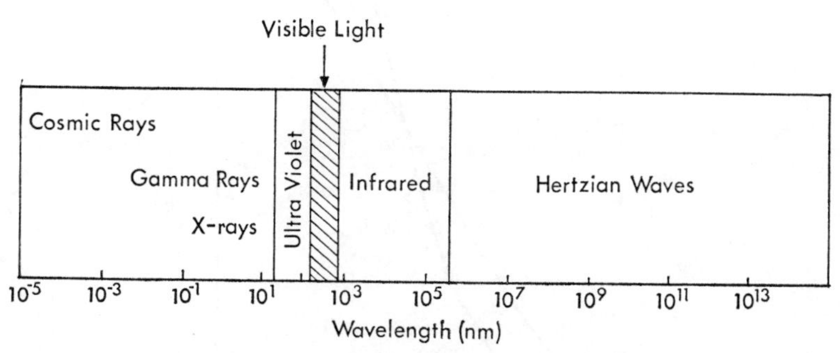

Fig. 1 The electromagnetic spectrum.

Wavelength and Frequency

The different kinds of energy in the electromagnetic spectrum may be described in terms of wavelength and frequency. We know that sound is propagated in the form of waves, with a certain distance between wave peaks. This is the wavelength which depends on the frequency of the waves.

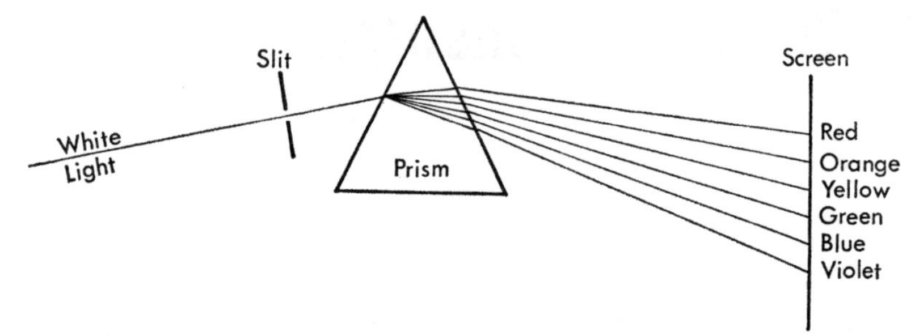

Fig. 2 Breaking down white light with a prism to form the color spectrum.

We know, too, that the higher the frequency, the higher the pitch of the sound.

Energy radiated in the frequency range from about 16 to 20,000 Hz (cycles per second) causes sound sensations to be created in the ear. The eye's sensitivity occurs at much higher frequencies. In this portion of the energy spectrum, it is more convenient to refer to wavelength rather than

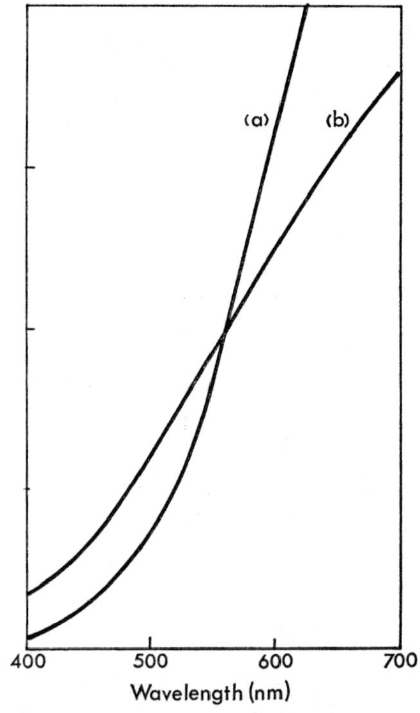

Fig. 3 Spectral energy distribution in the visible portion of the spectrum (a) curve representing the amounts of energy at different wavelengths for a tungsten lamp (b) effects on lamp emission of increasing the applied voltage.

frequency. Wavelengths in the range from about 700 to 400 nanometers (a nanometer is one-millionth of a millimeter) create visual sensations of color.

Light from a source such as a tungsten lamp radiates outwards in all directions in straight lines. If a narrow bundle of light rays is allowed to pass through a slit and fall on a prism, the strikingly beautiful color spectrum appears on a white screen or card placed at the opposite side of the prism. The paths of the light beams can be made visible by dust particles in the air, or by blowing smoke into the space at either side of the prism.

Figure 2 shows that the light rays responsible for the red sensation in the eye are bent the least in passing through the prism. These are the longer wavelengths, in the region of 700 nm. At the other end of the spectrum, in the blue-violet region, the shorter wavelengths—about 400 nm—are found. By placing a very narrow slit between the prism and the screen, single spectrum colors can be isolated, showing that each narrow band of energy produces a distinct color.

With the aid of sensitive measuring instruments, it can be shown that a tungsten lamp emits energy in a continuous band throughout the visible spectrum, as well as in the invisible infra-red and ultra-violet regions beyond the range of visual sensitivity. If the amount of energy from a tungsten lamp is measured, at say 10 nm intervals, it will be found that there is a great deal more energy in the longer wavelength region than at the blue-violet end of the spectrum. This is shown graphically in Fig. 3(a). Another very interesting experiment is to measure the effects of changing voltage or current in a tungsten lamp. When the lamp voltage is increased, more current flows, and the filament becomes hotter. The total amount of light from the lamp increases, and at the same time, the light takes on a whiter appearance. Fig. 3(b) shows that the amount of blue light relative to red increases as the voltage is raised.

Color Temperature and its Measurement

Color temperature is a term used to describe the spectral energy distribution of light sources. The color temperature scale in degrees Kelvin (K) refers to the actual temperature in degrees Centigrade of an idealized black body radiator plus 273 (absolute zero). A tungsten filament lamp emits light in a manner similar to a black body radiator, and its color temperature may be measured with a suitable instrument. In the laboratory the light from a lamp of unknown color temperature is matched in a visual comparison field with a black body radiator, the temperature of which is known. To make color temperature measurements in practical, every-day situations, meters have been designed indicating the ratio of blue to red energy in the light, and the meter dial is calibrated in K values. Meters of this type give accurate indications of color temperature only if the spectral energy distribution curves for the light sources being measured are similar to the ideal black body.

In practice, lamp emission may be modified considerably by reflectors, lenses, and accumulations of evaporated tungsten on the inner surfaces of the bulbs. Some color temperature meters make three measurements, by means of filters placed alternately in front of a photocell, to sample the light in three portions of the spectrum—red, green and blue.

The sun gives off light with a continuous spectrum, but when sunlight is compared with light from a tungsten lamp the latter appears to be quite yellow. Fig. 4(a) shows the spectral energy distribution of average sunlight.

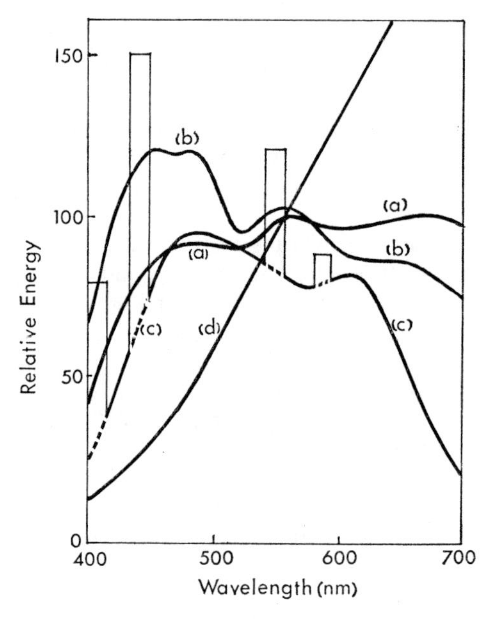

Fig. 4 Spectral energy distribution curves for (a) average sunlight (b) average daylight (c) fluorescent tube and (d) tungsten lamp.

When this curve is compared with the curve for a tungsten lamp (d), it can be seen quite readily that sunlight has a much higher energy distribution in the blue region of the spectrum.

Outdoor scenes are normally illuminated with a mixture of sunlight and skylight. An average daylight condition is shown by the curve in Fig. 4(b). Because these are not smooth curves, a meter measuring the blue-red ratio may give misleading indications of color temperature outdoors.

The spectral energy distribution curves for other types of light sources, such as arc lamps and fluorescent tubes, are as a rule even more irregular, to the extent that measurements of the blue-red ratio with a color temperature meter may be meaningless.

Some light sources—for instance, the mercury arc—give off energy in a number of intense spectral lines, with completely dark spaces between the lines.

Correlated Color Temperature

The light from a mercury arc may appear to the eye to be white in spite of the non-continuous characteristics of the radiation. In fact, the eye is unable to detect the difference between two sources with widely differing energy distributions, providing the appearance of the light is similar. The term correlated color temperature is often used to describe the whitish light from sources other than black body radiators. This term is defined as the color temperature of an idealized radiator producing light that most closely matches in appearance the source in question. For example, a warm white fluorescent lamp may be rated at a correlated color temperature of 3000K.

It is always very important to keep in mind that while this light may have a visual appearance similar to a tungsten lamp at the same color temperature, the spectral energy distributions are quite different, and the two light sources will produce entirely different effects on color film. This is due to the fact that the response of the film to color is not the same as that of the eye.

The correlated color temperature of average daylight is approximately 6500K. No known substance can be raised to this temperature by heating, to make an actual color match, but the spectral energy distribution of a full radiator, if it could be raised to 6500K, can be calculated from Planck's law. The distribution curves obtained in this way may then be simulated with the aid of color filters.

Trichromatic Color Vision

According to the theory of trichromatic color vision, any color can be matched by additively mixing suitable amounts of three colors, red, green and blue. (A few colors of high purity are exceptions to this rule, but a special technique may be employed to match these colors as well.) Assuming that the eye has three different types of response, R, G and B, it would not be able to detect a difference between two stimuli which give rise to the same R, G and B responses, no matter how different the spectral energy distributions may be. Fig. 5, showing the energy distributions of two white lights with similar appearance, demonstrates the theory of trichromatic matching.

A device for trichromatic matching, known as a colorimeter, is shown diagrammatically in Fig. 6. Any convenient set of three colors may be selected and mixed to illuminate the matching field. Red, green and blue filters may be utilized to obtain the three colors. Alternatively, using narrow slits, three bands of color may be isolated from the spectrum of a tungsten lamp. To achieve the highest purity in the mixture of colors in the matching field, three spectral lines may be selected from an arc source.

CIE System of Color Specification

A great number of different combinations of three colors will satisfy the matching requirements in a colorimeter, but for the results of colorimetric measurements to have universal significance, it is necessary to specify the reference stimuli in terms of which the results obtained with a particular colorimeter may be expressed. The amounts of the reference stimuli also must be specified.

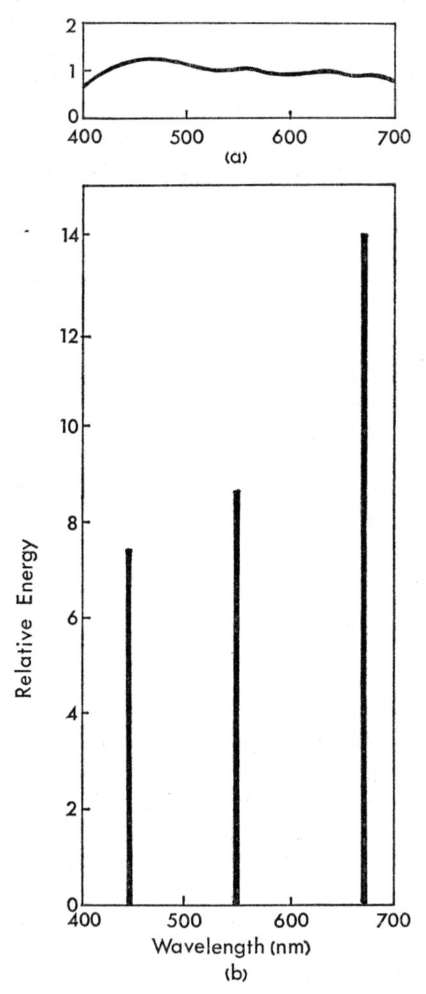

Fig. 5 (a) Spectral energy distribution of a source of white light (b) three narrow spectral bands additively mixed to match the appearance of the white light source.

In the CIE (Commission Internationale de l'Eclairage) system of trichromatic colorimetry three spectral lines selected from the mercury spectrum are used, at 700, 546.1 and 438.8 nm. The amounts of energy at these three wavelengths are specified in terms of an equi-energy spectrum—that is, one in which the energy per unit wavelength is the same throughout the visible spectrum. When the instrumental stimuli of any colorimeter are matched with mixtures of the three reference stimuli just described, their specification in terms of the three reference stimuli can be calculated readily, and in turn, it is possible to calculate the specification of any color measured with the colorimeter.

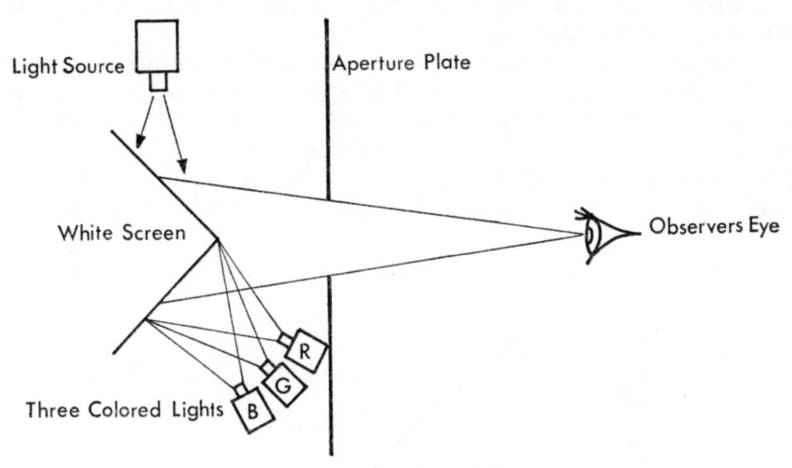

Fig. 6 Diagram of a colorimeter.

The color matching curves for the CIE standard observer are shown in Fig. 7. These curves were obtained by averaging the results of a number of measurements, in which persons with normal color vision participated.

The negative portions of these curves—the parts of the curves below

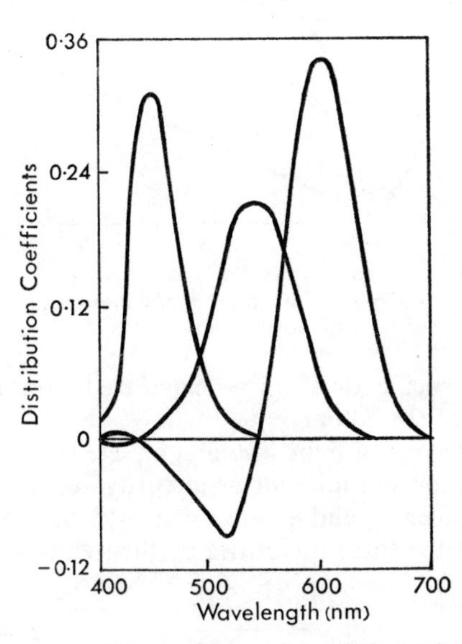

Fig. 7 Color matching functions of the CIE standard observer.

zero—require an explanation. When colors of very high purity, such as spectrum colors, are being matched with mixtures of three spectral lines, an exact match cannot be obtained in all parts of the spectrum unless a portion of one of the matching colors is added to the other side of the comparison field. This procedure may be illustrated in a convenient manner by negative quantities.

To avoid the complications of calculations involving negative quantities, CIE specified new stimuli X, Y and Z, with positive values only. These are not real lights, but mathematical abstractions by which any real light may be described, and are known as tristimulus values.

To obtain the XYZ specification of a color, it can be matched on a colorimeter, the results being modified by the appropriate transformation

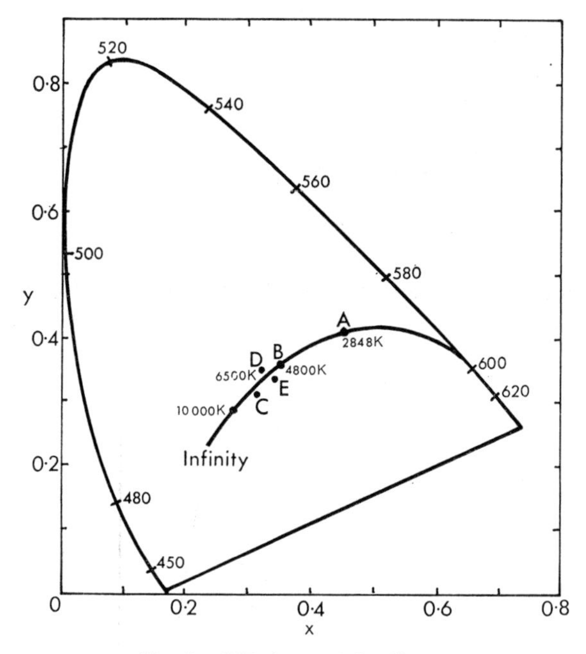

Fig. 8 CIE chromaticity diagram.

equations, or the specification can be calculated from the spectral energy distribution curve of the color.

The tristimulus values of light give a complete specification of that light in terms of both color and amount or intensity. To specify the color only, relative values are needed, and a convenient way of expressing the relation is to divide each of the three quantities by their sum—

$$x = \frac{X}{X + Y + Z} \qquad y = \frac{Y}{X + Y + Z} \qquad z = \frac{Z}{X + Y + Z}$$

These new values, xyz, are known as chromaticity coordinates, and are used in plotting colors on the CIE chromaticity diagram. Since $x + y + z = 1.0$, it is not necessary to plot all three values on the diagram, and the values of x and y have been selected for this purpose, as shown in Fig. 8. Equi-energy white with coordinates of $x = 0.333$ and $y = 0.333$ appears at the point marked E in the center of the diagram.

The curved line beginning at the point marked 700 at the extreme right of the diagram is the black body locus. On this curve the standard CIE illuminants, A, B and C are shown. This line does not pass through the point E because at no temperature does a black body give off energy with exactly the same color as that of equi-energy white.

Standard Illuminant A consists of a tungsten lamp at a color temperature of 2854K, while B and C are representative of sunlight and overcast sky respectively. These illuminants, standardized many years ago, are obtained with liquid color filters, to give color temperatures of approximately 4900 and 6700K. The increasing use of dyes and pigments which fluoresce has led to the standardization by CIE of a series of energy distributions representing daylight at all wavelengths between 300 and 830 nm. One of these representing standard daylight, has been designated D6500. The location of this illuminant on the CIE chromaticity diagram is shown in Fig. 8.

The horse-shoe shaped curve in the diagram represents the spectrum locus. Spectrum colors are located on this curve, commencing with red at 700 nm, and ending with blue at 400 nm at the left-hand corner. The straight line connecting these points represents mixtures of red and blue colors not found in the spectrum.

Response of the Eye

The color of the sensation is not the only way in which energy of one wavelength differs from another in its effect on the eye. There is also a quantitative difference—the relative luminous efficiency of radiation. Fig. 9 shows how the response of the eye changes throughout the spectrum, with maximum response at approximately 555 nm, in the yellow-green region. On either side of this point the response of the eye falls off rapidly as the limits of human vision are reached. It should be noted that this curve corresponds with the Y curve in the CIE color specification system.

It is not difficult to find the chromaticity coordinates of a light or a color from its spectral energy distribution. The converse is not true, however, because for every set of chromaticity coordinates an infinite number of spectral energy distributions can be found giving a match with the light in question when mixed in suitable proportions. These visually matching lights may give altogether different results when used to illuminate colored objects, or when a colored object illuminated by these lights is photographed with a color film.

Figure 5 shows the spectral energy distributions of two lights with identical visual appearance. Such stimuli, which are spectrally different but visually

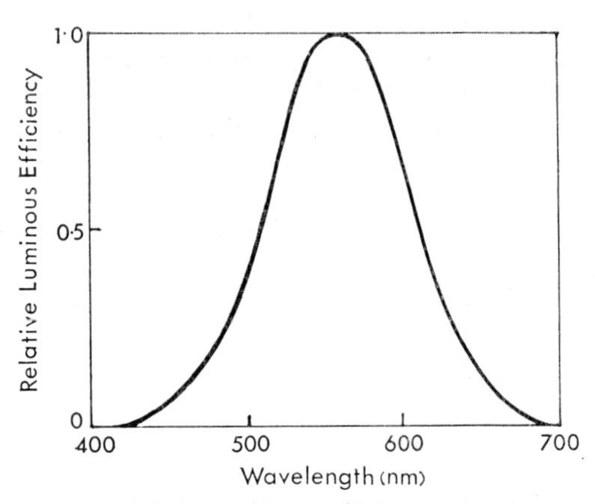

Fig. 9 Relative luminous efficiency of radiation.

identical are known as metameric pairs or metamers.

Many sources of light may be described as emitting white light. The sensation of whiteness is associated primarily with light in which all regions

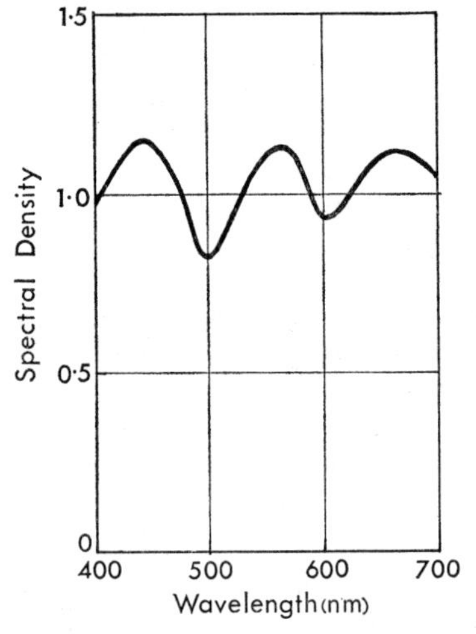

Fig. 10 Spectral density curve for a gray area in a color film.

of the spectrum are represented and in which no region is greatly in excess over any other region. Experiments with color mixing reveal that these statements are at best only broad generalizations, and comparisons of energy distributions for typical sources of white light, such as those in Fig. 4, show how widely the characteristics of these sources may vary.

When the energy distribution no longer conforms with these requirements, the light appears to be colored. The coloring power of a dye or pigment is nothing more than the power to reduce the energy in certain parts of the spectrum relative to the others. Measurement of the spectral energy distribution of the light reflected from colored surfaces shows that red surfaces, for example, reflect strongly in the region from 600 to 700 nm, but reflect very little energy in the remainder of the spectrum. White surfaces can be defined much more precisely than white light sources; a white surface is one which reflects energy uniformly throughout the spectrum. When one part of the spectrum is absorbed more than another, the surface becomes tinted. A gray surface is one in which the reflection factor is reduced at all wavelengths.

These considerations may be applied in a similar manner to transparent materials, such as motion picture film. Here the transmission factor of the material determines the color—and the intensity or amount—of the transmitted light. In the case of a color film, however, each of the dye layers has its own spectral energy distribution curve, and the transmission factor of the combination of three dye layers is much more difficult to define precisely. A neutral area in a color film might be specified by the mixture of the lights transmitted by the three dyes required to match a particular reference gray area. This reference gray might be obtained with one of the standard CIE illuminants at a suitably reduced level (see Fig. 10).

Additive Color Reproduction

The principle of additive color mixing described in previous paragraphs may be applied in practice to produce color pictures. Three projectors fitted with red, green and blue filters are needed, the three colored lights being superimposed on a reflecting screen in suitable amounts or intensities to obtain a visually neutral white. Simultaneously reducing the illumination from all three projectors in a uniform manner gives a range of grays, and a black screen results when all light is cut off. The illumination from projectors fitted with tungsten lamps cannot be reduced by simply lowering the lamp voltage or current, because that would alter the color temperature.

By cutting off one or more of the light beams, the six basic colors—red, green, blue, cyan, yellow and magenta—are obtained on the screen. For example, by cutting off the light from the red projector, a mixture of green and blue light results, giving the color cyan. This is the complementary color of red. When light from the blue projector is cut off, the remaining green and red lights are mixed to form yellow, the complementary color of blue. Similarly, the mixture of blue and red lights forms the color magenta when the green projector is cut off, magenta being complementary to green.

By reducing the lights from one or more projectors in varying amounts, a considerable range of other colors can be obtained. This may be done by interposing non-selective (neutral) light-absorbing filters in the light beams. To produce color pictures with the process, it is only necessary to place in the projectors three images on black and white film representing the amounts of red, green and blue light in the original scene.

The easiest way to demonstrate this process is to make use of a reversal type film which gives positive images directly, following exposure in a camera and subsequent development. (A positive image is one in which the lightest parts of the scene are represented by the lightest areas in the film image.) To obtain three images representing the red, green and blue light in the scene, three exposures must be made on separate films, using first a red, then a green and finally a blue filter over the camera lens. These images are known as separation positives.

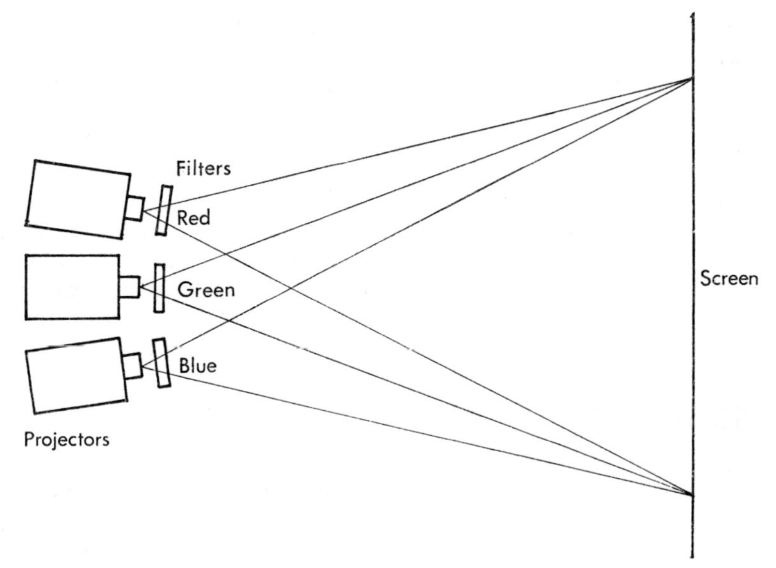

Fig. 11 The principle of additive color reproduction.

The positive made with the red filter is placed in the red projector, the 'green' positive in the green projector, and the 'blue' positive in the blue projector. When the three colored images have been carefully registered on the screen, a color picture is obtained, and if the process has been properly adjusted to allow the correct amounts of red, green and blue light to reach the screen, the color picture should be a reasonably accurate representation of the original scene.

Subtractive Color Systems

In additive methods of color reproduction colors are produced by mixing together suitable portions of three primary colors—red, green and blue. In motion picture applications these methods suffer from two major disadvantages. First, somewhere in the optical system there must be saturated red, green and blue filters, resulting in severe light losses. The second disadvantage is that special equipment is required, such as triple projection, with attendant losses in definition and registration problems.

The subtractive principle overcomes all these difficulties. While the manufacture and processing of subtractive film materials is much more complicated, the advantages of subtractive systems in motion picture applications far outweigh the disadvantages.

A typical color film consists of three light-sensitive layers responding selectively to the red, green and blue light in a scene. The upper layer is sensitive to blue light only, and exposure of the film produces an image in this layer representing all of the blue light in the scene. Under this layer an inert yellow filter layer is coated, preventing blue light reaching the remaining light-sensitive layers. Next is a layer sensitive to blue and green light only. Because no blue light can reach this layer, the image formed by exposure represents all of the green light in the scene. The bottom layer, next to the base, is sensitive to blue and red light only, and in this layer an image is formed representing all of the red light in the scene.

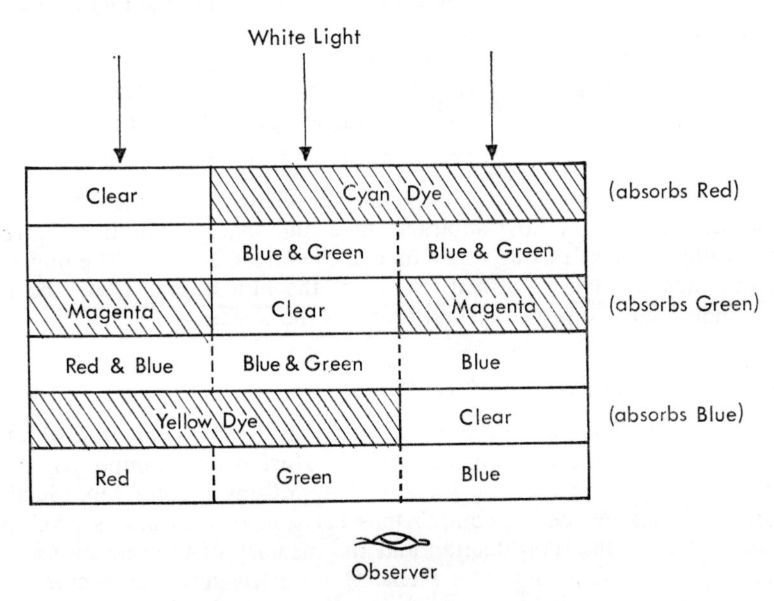

Fig. 12 The principle of subtractive color reproduction.

During development of the film, black silver images are formed in the three layers. In a later color processing stage, positive dye images are formed, a yellow dye image in the blue-sensitive layer, a magenta dye image in the green-sensitive layer, and a cyan dye image in the red-sensitive layer. When the silver images are removed by bleaching and fixing, only the superimposed dye images remain to form the color picture.

The dye colors are the complementary colors of the three additive primaries—red, green and blue. Cyan, for example, is complementary to red—that is, cyan is white light minus red. Similarly, magenta ls white minus green, and yellow is white minus blue.

Thus, the cyan dye layer in a color film freely transmits green and blue light while absorbing red; the magenta layer transmits red and blue, but absorbs green, while the yellow dye layer allows red and green light to pass through the film and absorbs blue.

As each of these dye layers transmits two-thirds of the while light and absorbs one-third, two dye layers super-imposed transmit only one-third of the light and absorb the remaining two-thirds, as the following table shows:

FILTER EFFECT OF COLOR FILM DYE LAYERS

Dye layer	*Effect*
Cyan plus magenta	Cyan subtracts red, magenta subtracts green; blue light passes through the two layers.
Cyan plus yellow	Cyan subtracts red, yellow subtracts blue; green light is transmitted.
Magenta plus yellow	Magenta subtracts green, yellow subtracts blue; red light comes through.

If all three dye layers are superimposed, all light is absorbed because cyan subtracts red, yellow subtracts blue and magenta subtracts green. Thus no light passes through the film and the screen is black. The opposite occurs when no dye is formed in any of the film layers. Then, the full while light from the projector lamp reaches the screen.

Visual Appreciation of Color

It is only rarely that a color reproduction can be directly compared with the original object or scene. Color memory therefore plays an important part. It might be assumed that the process of appraising a color reproduction consists of making mental comparisons between the sensations produced by the colors in the reproduction and the memory of the sensations produced by the original object or scene. Quite often, however, pictures are appraised by persons who have not seen the original colors. The basis of judgment is then a comparison between the color sensations aroused by the reproduction and a mental recollection of the color sensations produced by objects or scenes similar to those in the reproduction.

The color of the light illuminating an object or scene causes variations in hue. If the sun is low in the sky, for example, the light is yellowish. The apparent lightness of surface colors is affected by many factors. Any fabric that tends to hang in folds has troughs that look darker than the crests. Haze in the atmosphere can lighten dark objects and darken light objects if they are relatively distant. As with hue, so also with lightness, the background affects the appearance of objects. A dark background makes colors appear lighter, and a light background makes them darker. The apparent hue of an object is likely to change with the color of the background against which it is seen. Even the green of a tree is likely to be different if seen against a blue sky or a red building.

The saturation or vividness of colors as seen by the eye is likely to vary even more than hue or lightness. Scenes that appear to be dull and drab under an overcast sky become more colorful in direct sunlight, and the colors seem to have higher saturation.

Control of the iris of the eye is limited to about 8:1, so that a major portion of the adjustment in going from bright sunlight into a dimly lit room, for example, requires adjustment in the sensitivities of the different mechanisms of the retina. These changes in sensitivity, known as adaptation, are accompanied by marked changes in color vision, which result in colors appearing pale when intensity is low, and vivid when the intensity is high. An extreme example is the appearance of colors by moonlight, to the extent that they are often indistinguishable from grays.

From the foregoing it should be obvious that exact colorimetric fidelity is not necessary for color reproductions to be acceptable to the eye. In fact, some experiments have shown that exact reproduction of some colors such as flesh is not desirable. There is one property of the appearance of objects and scenes that remains remarkably constant, however, and that is the appearance of grays. A gray scale, seen in a wide range of conditions, continues to appear approximately gray. This is due partly to the ability of the eye to discount the color of the illuminant when viewing objects in its light. If two illuminants of markedly different color, such as tungsten light and daylight, are mixed in a scene, the result in a color reproduction is likely to be very unpleasant.

Color Terminology

It is generally agreed that colors have three main attributes—hue, which denotes whether the color appears to be red, green or blue, etc.; saturation, which describes the extent to which the color appears to be mixed with white, gray or black; and luminosity or lightness, the extent to which colors appear to give off or reflect more or less light. The appearance of a color can change with viewing conditions. For this reason it is helpful to divide color terms into subjective and objective designations.

Subjective terms denote the appearance of a color to an observer. Hue, saturation, luminosity and lightness are subjective terms. Objective terms refer to quantities that can be measured, and are therefore unaffected by changes in observer adaptation. These objective terms are correlated with the corresponding subjective terms as follows—

Subjective Term:	*Objective Term:*
Hue	Dominant Wavelength
Saturation	Purity
Luminosity	Luminance
Lightness	Luminance Factor
Hue and Saturation	Chromaticity

In the Munsell system of color specification* hue is used in the objective sense, value is used instead of luminance factor, and chroma instead of purity. The Munsell system has the special feature that, although an objective system, it is scaled subjectively, in the way that colors are seen by the eye. Equal steps of Munsell hue, chroma and value have been selected to represent as far as possible equal differences in hue, saturation and lightness. On a page of Munsell samples, where the columns of chips represent constant chroma, the saturation of the color increases with decreasing lightness, while in a series with constant hue and saturation, chroma decreases with decreasing lightness.

* Available from Munsell Color Co., 2441 North Calvert St., Baltimore, Maryland.

2 Color in the television system

Pictures are reproduced in the television system by means of a scanning process. In a monochrome television camera an optical image is formed by the lens on the face plate of the camera tube. Inside the tube an electron beam sweeps back and forth across the rear surface of the photo-sensitive face-plate, and electrical signals are generated which vary in amplitude in relation to the luminance variations in the optical image. These video signals are then mixed with synchronizing pulses, and impressed on a high-frequency carrier for transmission to the viewing public. At the receiver the television signal is picked up on an antenna, the video portion is extracted and fed to the picture display tube. The inner surface of the face plate of the picture tube is coated with a phosphor layer. An electron beam generated in the tube sweeps back and forth over the phosphor layer in a regular pattern of lines, and light is emitted, varying in brightness in relation to the amplitude of the incoming video signals. The action of the scanning beams in receivers is made to follow precisely the scanning beam in the camera at the television station where the program is originating by means of the synchronizing pulses transmitted with the picture signals.

Adding color to the television picture requires no major change in the television system itself, but special cameras and receivers are needed to generate and reproduce color. In the color camera three separate tubes are used to generate three video signals representing all the red, green and blue light in the scene. The picture tube in the receiver must have three separate electron guns to respond to the three color signals generated in the camera. The inner surface of the color picture tube is coated with large numbers of tiny red, green and blue phosphor dots, arranged in triads. A thin perforated metal screen located close to the phosphor layer ensures that the beam from the 'red' gun can strike only the red phosphor dots. Similarly, the 'green' and 'blue' beams reach only the green and blue dots.

An optical system in the color camera splits the image formed by the lens into three parts. (The fourth tube in some cameras generates luminance signals.) All of the red light from the scene is filtered out and directed towards the 'red' camera tube; all of the green light to the 'green' tube, and all of the blue light to the 'blue' tube. The signals generated in the camera tubes represent all of the red, green and blue light in the original scene, and the amplitude of these signals varies, depending on how much of each color is present in each part of the scene, as the scanning beams sweep back and forth across the images. If, for example, a scene includes a man in a bright red uniform, the signal from the 'red' tube would have high amplitude in this part of the picture, while at the same time the green and blue signals would have much lower amplitude.

Monitoring Color Television Signals

Very few colors in nature are pure. Only rarely will an object be found reflecting just red light and absorbing all of the green and blue light. When a scene is televised with a color camera, signals are present at the camera output in relation to the amounts of red, green and blue light reflected by the objects in the scene. These signals are displayed in the camera control unit on a waveform monitor, a special type of oscilloscope designed for this purpose. Attached to the face of the waveform monitor tube is a graticule with a scale marked off in IEEE units. When the signals from a color camera are displayed on the waveform monitor, the amplitude of the signals may be measured relative to the graticule scale. On this scale zero represents absolute black, while 100 indicates white level.

With no light entering the camera lens, signal output is at a minimum, corresponding to black. The traces on the waveform monitor representing this level are adjusted to coincide with zero on the graticule scale. Now, as light is allowed to enter the camera lens, the amplitude of the signals increases until a point is reached at which the lightest parts of the scene—whites or light colors—reach 100 on the scale. This is the maximum level that can be transmitted in the television system. If the signal levels increase beyond 100, the camera lens aperture must be closed down, the lighting on the scene must be cut back, or the gain of the signal generating system must be reduced (see Plate 1).

Color Camera Alignment

A basic requirement in color television is that neutral objects—whites, grays and blacks—must be reproduced at the same signal levels in the three color channels. It is customary in setting up color cameras to utilize a gray scale chart or a light box fitted with a neutral step wedge to establish uniform levels in the signals displayed on the waveform monitor. When a camera set up in this way is directed towards a scene, pictures with a

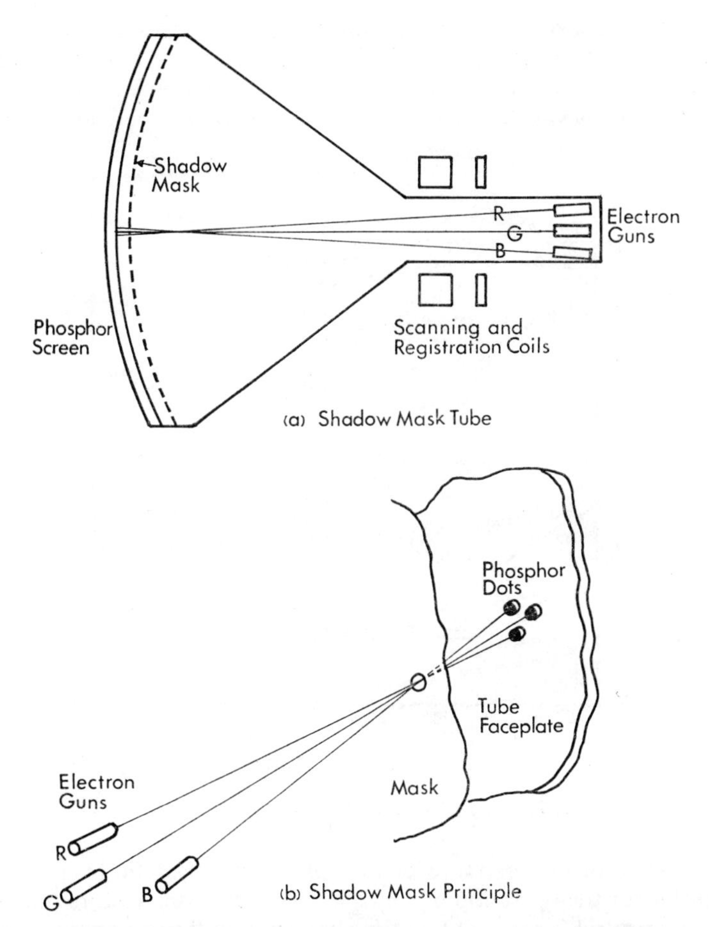

Fig. 13 Diagrammatic representation of a color picture tube.

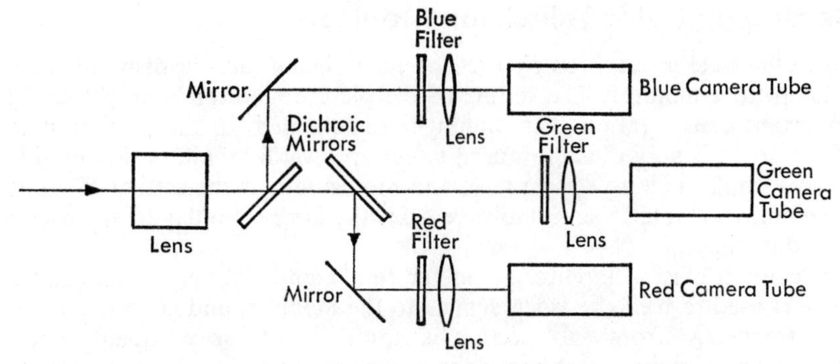

Fig. 14 Essential components of a color television camera.

pleasing appearance should be obtained. In some cases the program director may not be entirely satisfied with the pictures, and it may be desirable to alter face tones slightly, or change the color balance by a small amount, to achieve artistic effects. This can be done by making small adjustments in

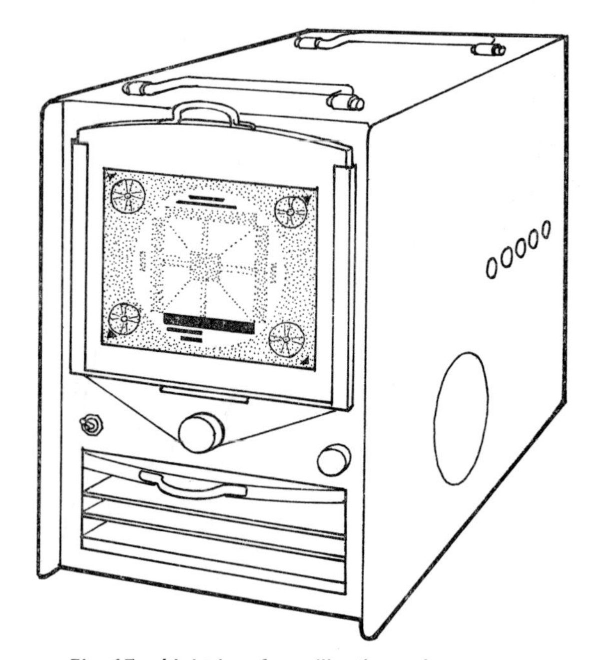

Fig. 15 Light box for calibrating color cameras.

the amplitude of one or more of the color channels. The technical crew responsible for the operation of the equipment can easily restore the system to normal neutral balance at any time by directing the camera towards the calibrating gray scale or light box.

Setting up Color Television Monitors

To evaluate color television pictures the video signals must be displayed on a color picture monitor. The setting of the picture monitor is an extremely important consideration that cannot be disregarded. In the present state of the art, it is very difficult indeed to set up several monitors side-by-side to give similar picture appearance, and to ensure that the setting of all of the monitors in a television station remains the same from day to day over a period of time.

A color television monitor is similar to a home receiver, except that a cable is used to feed the video signals to the monitor, and consequently a high frequency 'front end' section is not required. Better quality components and circuitry must be used in monitors to provide the best possible pictures. The studio color monitor is the reference and provides the primary color standard by which the operation of the entire system is judged.

There is more or less general agreement throughout the television industry that color picture monitors should be adjusted to reproduce white at a color temperature of D6500. Originally, the National Television Systems Committee (NTSC) established reference white at Illuminant C ($x = 0.310$, $y = 0.316$) which is about 6800K. Recently Illuminant C has been replaced by a new designation for average daylight known as D6500 (see p. 9).

Television Reference White

In practice it is very difficult to adjust color monitors accurately to this white condition. Color is produced in a television picture tube by mixing three colored lights—it is an additive system. The amounts of red, green and blue light can be changed easily by raising or lowering the signal levels applied to the three guns in the picture tube.

When a white card is placed in front of a television camera, the signal levels from the three camera tubes can be adjusted to have equal amplitudes as observed on the waveform monitor at the camera control unit. Now if these signals are applied to a picture tube, it is unlikely that the area on the face of the tube representing the white card will have the same white color as the card itself. This is due to the fact that the three phosphors used in the manufacture of the picture tube screen do not emit light equally for equal input signal levels. By means of individual gain controls in the three color channels the signal levels may be raised or lowered to obtain any desired shade of white in the area on the tube where the image of the card appears.

Monitor Calibration Methods

Many different methods have been devised to set up color monitors in a uniform manner. Some of these methods depend on visual comparison, while various types of instruments are used in others. The simplest method is to set picture monitor white to match an illuminated surface in a reference light box. To make the match, the light box is placed close to the picture monitor, and the color gain controls on the monitor are trimmed slightly one way or another to produce a white area on the monitor as close as possible in appearance to the illuminated reference surface in the light box. Usually the red gain of the monitor remains fixed, while the green and blue gain controls are adjusted.

With this method the calibration of the reference light box is the critical factor. Light boxes from different manufacturers seldom give a good match among themselves, and the lamps for illuminating the reference surface are subject to change with age and applied voltage.

There are many different types of color temperature meters on the market, and some of these have been used for setting up color picture monitors. However, a meter designed to measure the color temperature of continuous sources of radiation (i.e., tungsten lamps) is likely to give erroneous readings when used to measure a white area on a color picture monitor, where the light is a mixture of three colors (see p. 3).

A color calibrator for monitors in television studios has been designed by the National Research Council, Division of Applied Physics, Ottawa, Canada.* In this device a reference light source is provided by a quartz iodide lamp illuminating a diffuse surface which is seen in half of a circular

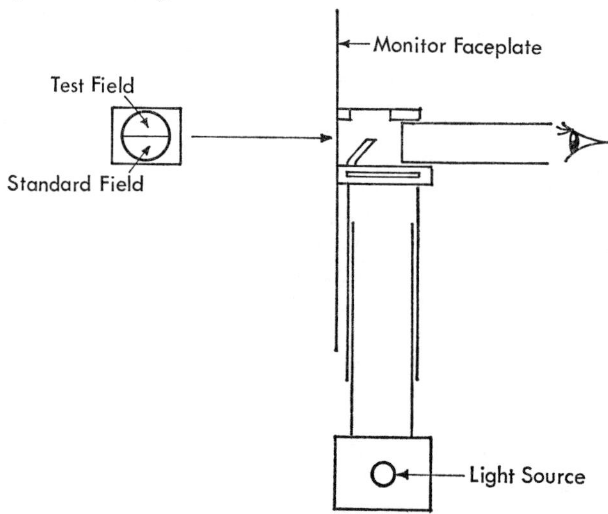

Fig. 16 Calibrating device for color picture monitors.

viewing field. The other half is diffusely illuminated by a white patch on the picture monitor tube. The reference illumination is modified to the same chromaticity as D6500, and the color gain controls of the picture monitor are adjusted to obtain the best visual match between the two halves of the viewing field. By means of an attenuator, measurements may be made at two levels of monitor brightness. In this way the neutrality of the picture monitor throughout the gray scale (tracking) may be checked.

A paper by M. Rotthaler in the German publication, *Fernseh und Kino-Technik*, August 1969, with the title "Achromatic Trimming of Color Picture Tubes" surveys the problems of color monitor adjustment and describes two instruments developed by Institut für Rundfunktechnik in Munich. In this paper the question of required accuracy of adjustment is also considered.

Television Scanning Systems

All television systems use the horizontal scanning principle in which an electron beam is made to traverse the picture area (raster) from left to

* A paper describing this color calibrator appeared in the June 1968 issue of the *SMPTE Journal*, p. 622.

right, commencing at the top of the picture. To minimize flicker, the beam is made to scan alternate lines in a pattern of interlaced fields. First, all of the odd lines are scanned—1, 3, 5, 7 and so on. Then the beam returns to the top of the picture to begin the scanning of the even lines making up field 2. The lines from these two interlaced fields make up one complete picture frame.

In the North American system there are 525 lines in a frame, and frames are repeated at the rate of 30 per second. The European CCIR (International Radio Consultative Committee) system has 625 lines per frame and 50 fields (25 frames) per second.

Color Transmission Systems

A basic requirement in all color television systems is compatibility—that is, color signals must produce pictures on monochrome receivers with no need for modification of the receiver. A color television signal must have, then, all of the characteristics of a monochrome signal, and in addition, information which will give rise to color pictures on color receivers. This rules out at the start any scheme of color television transmission requiring separate red, green and blue signals to be sent directly to receivers. Such a scheme would be impractical in any event since it would require a space in the radio frequency spectrum three times as wide as monochrome television.

There are three different systems of color transmission in use at the present time—NTSC, PAL and SECAM. The NTSC system was developed in the United States; PAL originated in Germany and was adopted later on in the United Kingdom, while SECAM is a French innovation. All of these systems start and end with the three primary signals, red, green and blue. A slightly different method of sending color information is employed in each system.

NTSC System

In the early 1950's experiments were conducted in the United States to find a solution of the problem of transmitting color pictures in the normal monochrome television channels. Studies of the behavior of the human eye indicated that the structures concerned with perception of color and brightness differ considerably in their sensitivity, especially in respect to color detail. The eye's ability to distinguish fine detail in color is, in fact, negligible. These studies demonstrated that color information could be transmitted with much narrower frequency bands than brightness information.

An entirely new concept was developed in which the output signals from the three tubes of the color camera are coded by a process of linear combination to produce separate luminance and chrominance signals. By adding together 30 per cent of the red signal, 59 per cent of the green signal and 11 per cent of the blue signal, a new signal is obtained representing the brightness of the scene. When this new signal is fed into a monochrome monitor or receiver, the picture has an appearance similar to that obtained with a single-tube monochrome camera. This is the luminance (Y) signal.

To obtain the chrominance signal, the luminance signal is subtracted from the three camera signals. This results in three signals representing red minus luminance (R–Y), green minus luminance (G–Y), and blue minus luminance (B–Y).

The signal G–Y is superfluous, because it can be reconstituted in the color receiver by combining appropriate proportions of R–Y and B–Y. These two difference signals could be transmitted along with the luminance signal and produce color pictures on monitors and receivers, but better

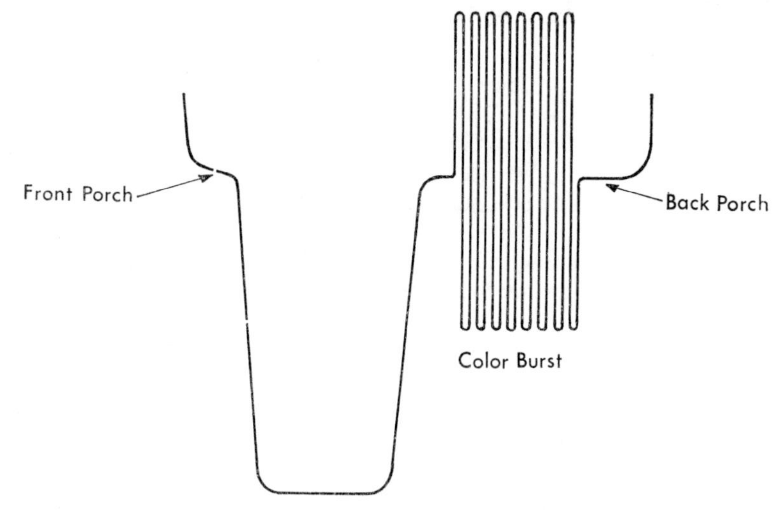

Fig. 17 Horizontal synchronizing pulse and color burst.

performance of the system is achieved by again combining certain proportions of the difference signals to obtain what are known as the I and Q signals.

A matrix unit or encoder is utilized to derive the luminance and chrominance signals from the outputs of the red, green and blue tubes in the color camera. By a process of subcarrier modulation and frequency interleaving, the chrominance signals are mixed with the luminance signal. In color receivers a decoder recovers the original red, green and blue signals and these are applied to the guns in the color tube to produce color pictures, as previously explained. In monochrome receivers only the luminance signal is effective, the chrominance information being rejected.

A few cycles of a color synchronizing signal at a frequency of 3.58 MHz are transmitted at the beginning of each line of picture information. This is known as the color burst, shown along with a horizontal synchronizing pulse in Fig. 17. The color burst provides a reference for recovering the color signals in receivers.

Color reproduction in the NTSC system is phase-dependent—that is, the phase relationship of the modulated subcarriers to the reference frequency must be maintained accurately. Fig. 18 is a color phase diagram, showing the correct angular displacement of the I and Q signals, the color difference signals, and the six primary and complementary colors, relative to the phase reference.

SECAM System

The NTSC system has been criticized for its susceptibility to phase distortions of various kinds, and many proposals have been put forward for color modulating methods that would overcome this drawback. One of these is SECAM (*SEQU*ential *A* *M*emoire) developed by Compagnie Francaise de Television, in Paris. In this system only one subcarrier is used, and this subcarrier is modulated sequentially by the two color difference signals. There is no need to transmit a color burst with SECAM, but a synchronizing signal is included at the end of each field to control the phase of an electronic switch in monitors and receivers. By means of a device known as a delay line, one of the two color difference signals is delayed for an interval equal to one line scan, so that both are made available simultaneously to form the color pictures in monitors and receivers.

PAL System

The PAL system (*P*hase *A*lternation *L*ine) was developed in the Federal Republic of Germany by Telefunken AG. and has been adopted in Great Britain as well. PAL encoding is similar to NTSC except that a switch is included reversing the phase of the subcarrier during successive lines. When a phase error between chrominance and burst signals occurs, the hue change introduced during one line is followed by an equal and opposite hue error in the succeeding line. This tends to average out the effects of phase errors.

In addition to switching pulses to operate a phase-reversing arrangement in the receiver, color burst must be transmitted to control the frequency of the receiver oscillator. A disadvantage of the PAL system is the increased cost and complexity of receivers.

Color Bar Calibration Signal

Additive color systems have three primary colors—red, green and blue—and three complementary colors—yellow, cyan and magenta, which are obtained by mixing pairs of primaries: red-green, blue-green and red-blue. White is obtained by mixing all three primaries and black where no light is present.

In the television system these colors are converted into signal voltages which can be measured, and displayed on a waveform monitor.

A standard color bar test signal simplifies the checking and adjustment of the television system. This signal is generated electronically, and consists of seven equal intervals in descending order of luminance—gray, yellow, cyan, green, magenta, red and blue. These signal levels correspond to saturated colors transmitted at 75 per cent of full amplitude, occupying three-quarters of the active scanning lines. The remainder of the signal is taken up with special test information including white and black reference

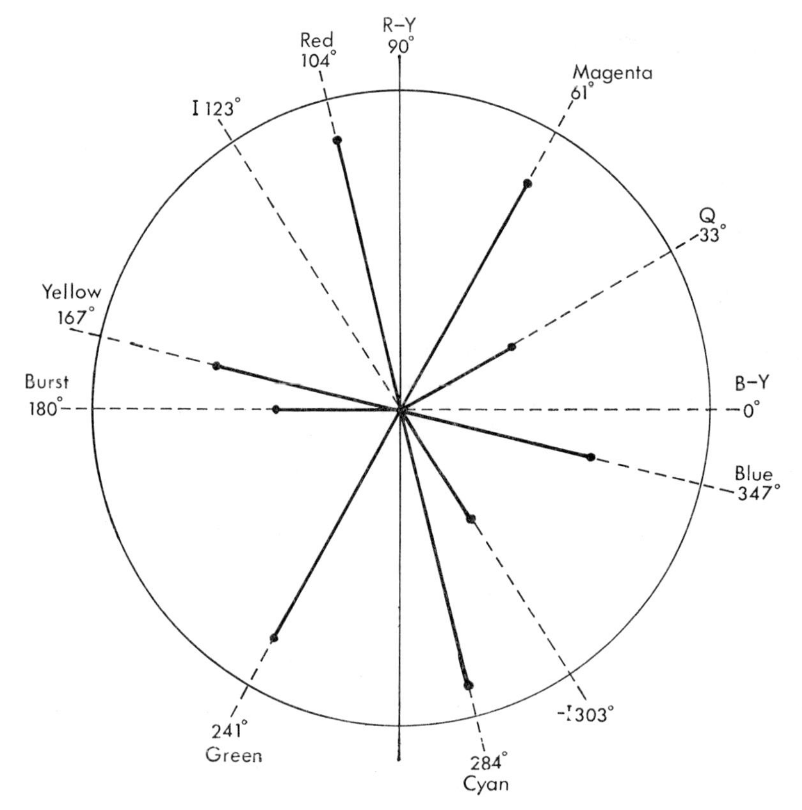

Fig. 18 Color phase diagram.

levels. On a picture monitor the test signal is displayed in the form of vertical bars, with the white and black references in a space at the bottom of the picture area.

The color bar test signal can be used to adjust encoders and color monitors and to check transmission systems. With a special type of test instrument known as a vectorscope, the pattern produced by the signal can be evaluated in relation to an engraved scale on the face of the vectorscope tube. This scale is derived from the color phase diagram shown in Fig. 18. A correctly adjusted encoder produces a pattern as shown in Plate 2.

The color bar signal is useful also in making phase and decoder adjustments in color monitors. With the red and green guns switched off the monitor phase control is adjusted to give equal brightness in the remaining blue bars. When the green and blue guns are turned off the two red bars should be equally bright.

Sources of Television Programs

At one time all television programs had to be produced at the time of the scheduled transmissions to the public. These so-called 'live' programs were supplemented heavily with motion picture films. In 1956 a method

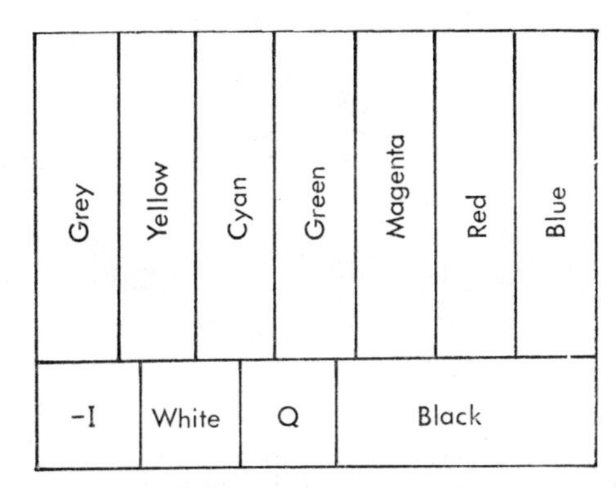

Fig. 19 Pattern produced on a color monitor by the color bar signal.

of recording television signals on magnetic tape was announced, and many broadcasters anticipated that all-electronic systems would eventually displace film altogether.

Video tape recording has led to pronounced changes in television programming practices, but it has had little if any effect on the extent of film use. In fact, the use of film has become firmly established, many of the most popular programs being made on film especially for television. However, there is very little 'live' programming nowadays. As a rule programs are pre-recorded on video tape, and new techniques of program production have been developed to take advantage of the video tape medium.

Video Tape Recording

Techniques for recording electrical signals on magnetic tape have been well established for many years, notably in radio broadcasting where extensive use is made of ¼-in. audio tape. In this application the frequency range to be recorded is about 15,000 Hz. This can be achieved easily at a tape speed of 7½ in. per second. At this rate the highest recorded frequency has a wavelength of ½ mil—that is, 1/500 of an inch.

Television signals have a frequency range of 4.2 MHz, or about 280 times greater than audio signals. To record this frequency range, retaining a wavelength for the highest frequency of $\frac{1}{2}$ mil, the speed of the magnetic tape over the recording head would have to be increased to about 175 ft per second, or nearly 2 miles per minute. If a recorded wavelength of 1/5 mil could be tolerated, tape speed could be reduced to 800 in. per second, but this would still require a reel of tape 120,000 ft in length to record a half-hour program.

Intensive research in the mid-1950's demonstrated the feasibility of a new recording concept in which a rotating head wheel recorded patterns across the width of a magnetic tape 2 in. in width, while the tape was being advanced at the rate of 15 in. per second. This was accomplished with a head wheel rotating at 14,400 rpm. The head wheel had four equally spaced

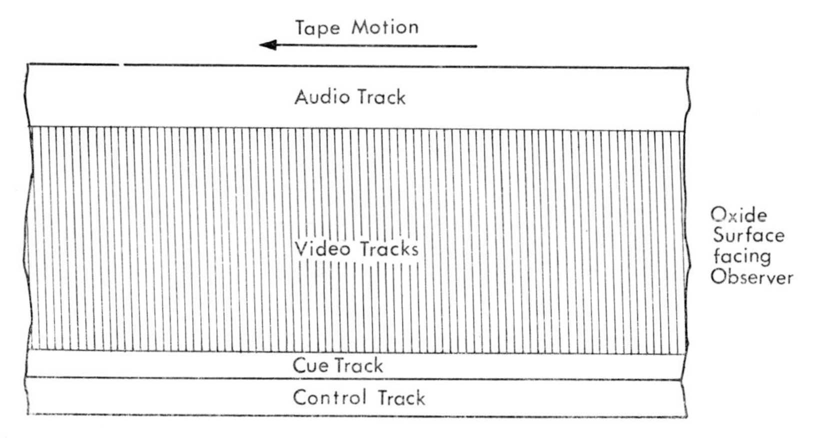

Fig. 20 Recorded pattern produced by rotating head wheel on video tape.

heads around its outer edge. This equipment produced so-called quadruplex video recordings. Four rotations of the head wheel recording 16 video tracks across the tape made up one television field, including vertical synchronizing information.

The machines used for video recording are quite large and complex. A roll of tape 2400 ft in length weighing about 20 lb is required for a half-hour program. The tape transport mechanism must move the tape at a constant rate through the head wheel assembly to lay down an accurately spaced pattern of recorded tracks. A vacuum tape guide forms the moving tape into a concave configuration as it passes the rotating head wheel to ensure intimate contact with the oxide surface across the entire width. At the same time audio, control and cue tracks are recorded longitudinally on the edges of the tape.

In the playback mode, a servo-mechanism adjusts tape speed and positioning to ensure that the rotating head traces out the originally recorded pattern on the tape. The entire composite video signal including the synchronizing information is recorded and must be recovered in playback with minimum distortion and time displacement, relative to a reference synchronizing generator. Minor errors which would not affect appreciably the picture portion of the signal may seriously interfere with the recovery and transmission of the recorded signals.

From the beginning, video tape recording was an unqualified success, and soon most television stations had at least one recorder. While the original objective in designing video tape equipment was to provide a high-quality recording and playback system for storing and delaying complete programs, broadcasters began almost at once to explore the possibilities of utilizing the equipment in program production.

The simplest production requirement—joining together two scenes recorded at different times—could be achieved by physically splicing the tapes at the desired points. These crude methods were soon replaced by electronic editing techniques, and highly sophisticated equipment was developed for this purpose. Now programs can be assembled on tape in a variety of different ways.

A popular technique with many program producers is to transfer film to video tape, to take advantage of electronic editing and special effects, avoiding at the same time the physical cutting and splicing of the film.

Another interesting technique makes use of A & B film rolls, mixed together on to video tape from telecine, using two synchronously running telecine machines.

Film as a Source of Television Signals

In the earlier monochrome television era, equipment for telecasting motion picture film consisted of a film projector set up in front of a television camera, with the projector lens forming a sharply focused image on the face plate of the television camera tube. In North America, the vidicon tube came into general use for this purpose, and the term 'telecine' was coined to describe an assemblage of equipment for telecasting film.

Basically, signals are generated in a color telecine camera in much the same way as a color studio camera, except that the scene in front of the studio camera is replaced by film pictures in the projector gate. The optical system in the camera splits the light beam from the projector into red, green and blue components, and sharply focused images of the pictures in the projector gate are formed at the face plates of the camera tubes.

In England and Europe, and some other countries as well, where the CCIR 50-field, 25-frame television system is in use, flying spot scanners are very popular, especially for the 35 mm film format. Signals are generated in flying spot scanners in an entirely different manner, compared with vidicon telecines. This subject is dealt with in greater detail in Chapter 3.

3 Reproducing color film

Basically, all that is needed to reproduce film in the television system is a device to project the film pictures into a camera, while at the same time the sound track is scanned with a light beam and photocell. The light patterns formed by the projector lens on the face of the camera tube have characteristics similar to the images projected by the lens when a studio camera is directed towards an actual scene, and the television scanning process takes place in exactly the same way in both cases.

Television equipment designers have developed a number of different types of machines and systems for reproducing film. A popular telecine layout used extensively in North America consists of two film projectors, a 2 × 2 slide projector, a television camera and an optical multiplexer, as shown in Fig. 21. In Europe, flying spot scanners are favored for reproducing film. In this type of equipment the light source is a cathode ray tube. The moving spot of light produced on the face of the tube by a scanning beam is imaged by a lens at the plane of the film. Light passing through the film to a photo-electric cell generates the video signals.

The difference between European and North American television frame rates has been responsible for the development of these different systems of film reproduction. In Europe, it is the usual practice to reproduce film at the rate of 25 frames per second, coinciding with the television frame rate. In North America it is impractical to increase the rate of film movement to 30 frames per second, and the practice has been adopted of reproducing the film at the standard motion picture frame rate of 24 per second, while a storage-type camera tube is used to generate the video signals.

Another important difference between European and North American practice is the method of film transport. Continuous-motion projectors are favored in flying spot scanning systems, while camera-type telecines normally use intermittent projection equipment.

Camera-Type Telecines

In the equipment layout shown in Fig. 21, the light beam from any of the projectors may be selected and directed into the camera through the multiplexer. A field lens is used in the extended optical system, the lenses of the

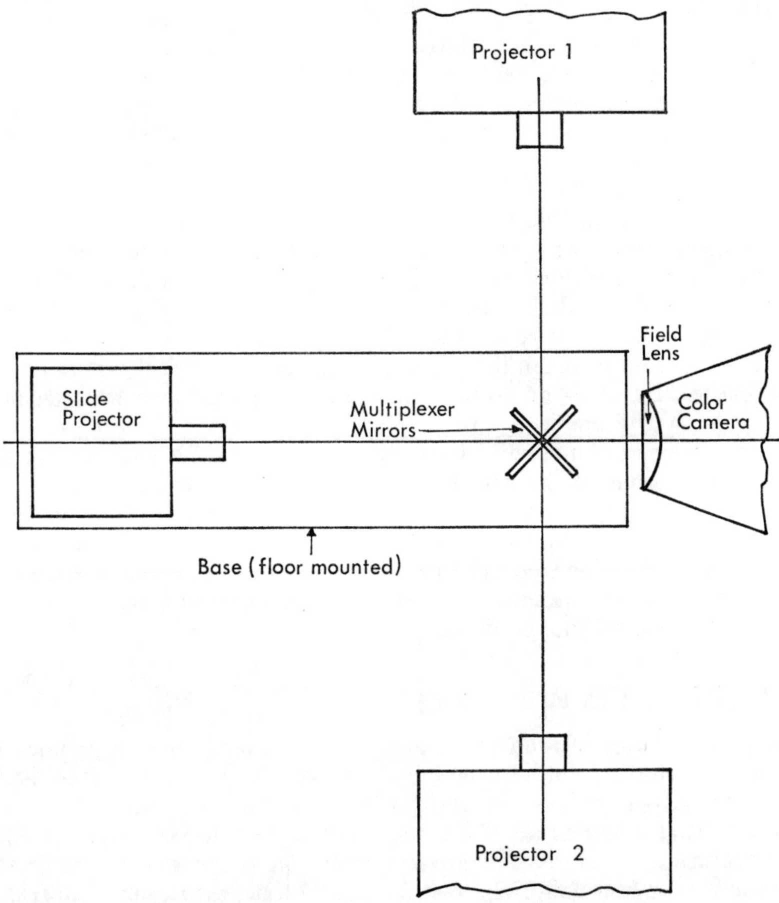

Fig. 21 Light paths in a multiplexed telecine.

projectors and the television camera being focused on a plane at the field lens.

Projectors specially designed for telecine service are a far cry from the inexpensive, portable machines in common use for showing 16 mm films on screens. Television film projectors must be ruggedly constructed to withstand continuous on-air service, with a high degree of reliability. Some of these projectors have provision for forward and reverse film movement, an important advantage in television operations. Other highly desirable features are fast wind and rewind transport, still-frame projection, and automatic cueing.

Slide projectors for television service are usually controlled by a remotely-actuated motor drive. It should be possible to register the slides accurately in the changer mechanism, in relation to the television raster, to ensure that titles and other pictorial material can be properly centered and aligned.

There are many different methods of multiplexing the light beams from the three projectors into the television camera. Basically, a multiplexer is a precision optical bench, with the projector light beams directed into an assembly of mirrors or prisms, from which the selection of the desired projector light beam may be effected. A popular arrangement consists of two front-surfaced mirrors, which may be shifted vertically by air pressure, remotely controlled from the telecine operating position.

Color cameras used in telecine installations are similar in many respects to studio cameras. Vidicon cameras, used extensively for telecine service in the earlier monochrone era, could be mounted directly on the multiplexer, considerably simplifying optical alignment. Color cameras, being much larger and heavier, must be separately floor-mounted, and lined up accurately with the light beams from the projectors through the optical system of the multiplexer. Either 3- or 4-tube cameras may be used in color telecines, with vidicon or Plumbicon tubes.

As in the operation of television studio cameras, the video signals appearing at the output of the camera must be monitored and processed in a camera control unit, before being fed into the television transmission and distribution system. Many different methods of telecine camera control are in use in television stations. In a station with more than one telecine it is customary to group the camera control units in a central area, enabling an operator to control the outputs of several telecines.

Television Film Projectors

The film projector in a color telecine performs exactly the same function as the projector in a monochrome installation. The physical characteristics of color film—thickness, dimensions, etc.—are identical with black and white films. The same rate of film movement applies to both types of film, and registration requirements are the same. The purpose of the projector is to move the film uniformly, one frame at a time, registering each frame accurately in the gate, where optical image formation takes place.

A number of different methods may be employed to move the film intermittently, but the most popular method consists of a toothed claw mechanism engaging the perforations, and being moved vertically by a rotating cam mechanism. Above and below the gate, the film is formed into loops by continuously-moving feed and take-up sprockets. These loops absorb the intermittent motion of the film in the gate.

The film gate consists of an aperture plate with a fixed guide, and spring-loaded pressure shoes to ensure accurate horizontal positioning of the film relative to the aperture. The opening in the aperture plate is machined to conform with the standard motion picture frame size.

Numerous methods have been employed to project films taken at the standard 24 frames per second rate into the television system in which the frame rate is 30 per second. Intermittent frame-by-frame projection requires a storage type tube such as the vidicon to be used to allow portions of television fields to be blanked out while the film is being moved in the gate.

Some 16 mm projectors make use of a film-advance mechanism with less than 50 degrees of shaft rotation for each film frame, two shutter

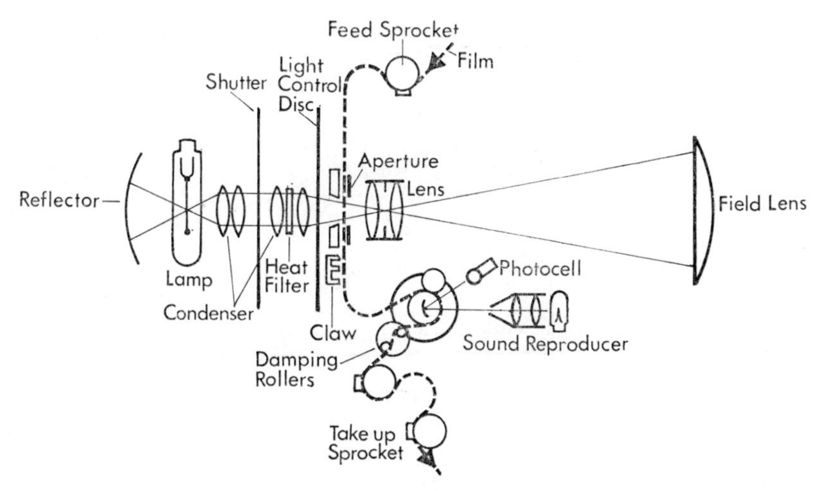

Fig. 22 Film path and gate assembly in a telecine projector.

interruptions per field (120 per second), and a uniform pull-down rate of 24 per second, or once every $2\frac{1}{2}$ television fields, to achieve the desirable 30 per cent light application time. Another arrangement uses a 3:2 system of film advance. With this method, film advance is initiated at alternate intervals of time corresponding to three and two television fields—that is, at alternate time intervals of 1/20 and 1/30 sec, or a total time of 1/12 sec for each two consecutive frames, giving the correct average rate of 24 per second.

Color Telecine Camera

The basic function of the telecine camera is to convert colored optical images into video output signals, which, after appropriate modification and processing in the camera control unit, can be transmitted in the normal manner.

The light beam from a projector, reflected from the multiplexer mirror, enters the television camera through the field lens. The projector lens forms a real image at the plane of the field lens. This image may be made visible by means of a white plastic test plate that can be lowered in front of the field lens. In this way, each projector lens can be sharply focused, using suitable test objects.

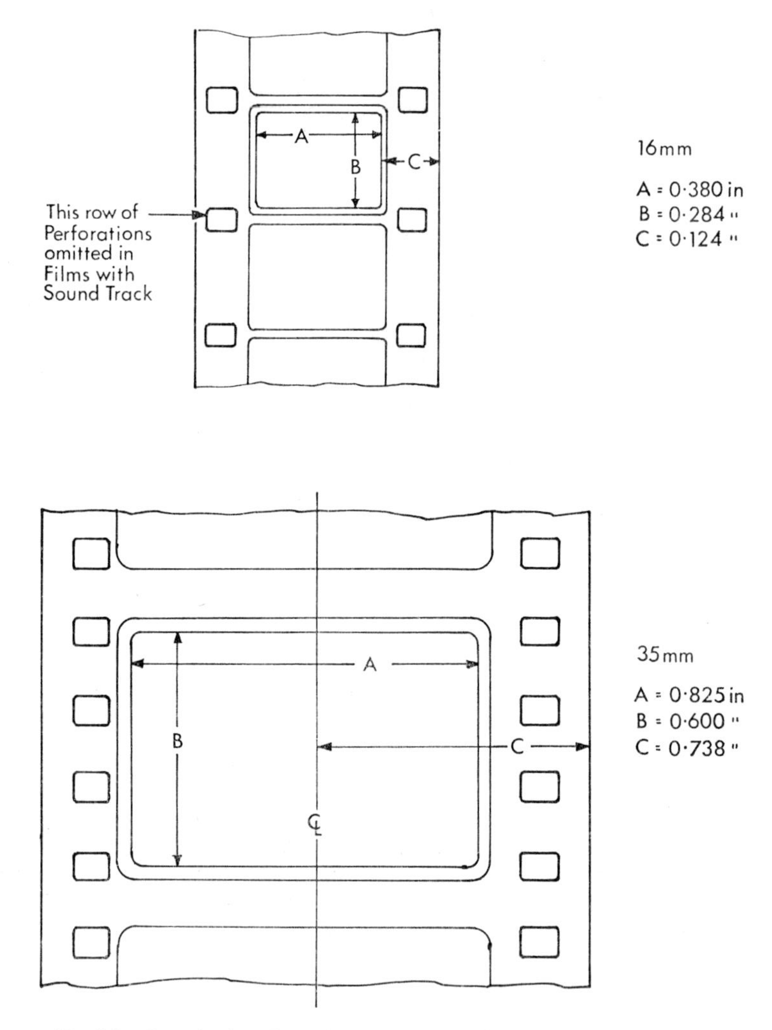

Fig. 23 Standard projected image areas of 35 mm and 16 mm films.

The standard EIA resolution slide is shown in Plate 3(a). This slide may be used for focusing the lens of the slide projector. The resolution and alignment section of the SMPTE television test film shown in Plate 3(b) is used for checking and adjusting the focus and image positioning of the film projectors.*

Correct image size from the film and slide projectors can be established at the same time by positioning the arrow points in the projected images to coincide with lines engraved on the test plate. These arrow points represent the areas of the film images scanned in the television camera.

A number of different methods are used to divide the light entering the camera into its red, green and blue components. Some cameras have four tubes, one providing a luminance signal. In a four-tube camera, a beam-splitting prism immediately behind the field lens divides the light beam into two equal segments, one to provide the luminance and the other the

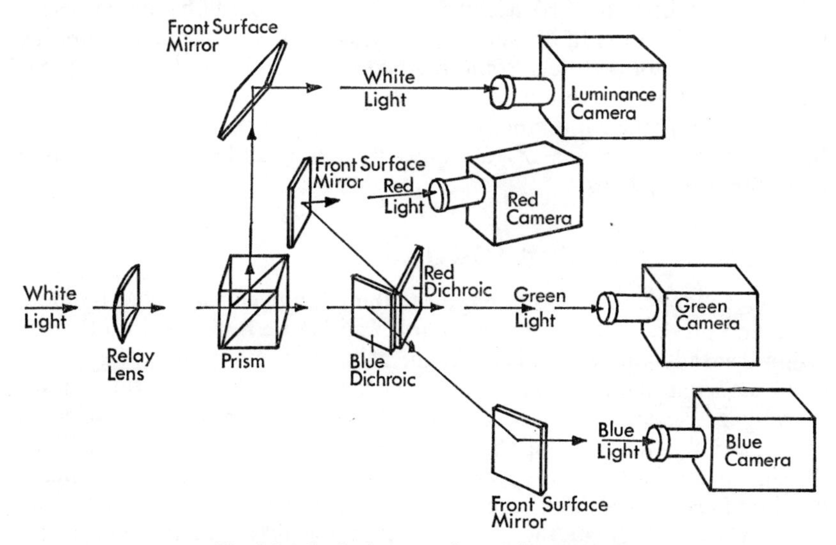

Fig. 24 Optical system in a color camera.

chrominance signals. A front-surfaced mirror directs the luminance segment to a vidicon tube, a trimming filter shaping the spectral energy distribution of this light beam to match the sensitivity curve of the human eye.

The chrominance segment of the light from the beam-splitter falls on a blue dichroic mirror that transmits red and green light but reflects blue light. The blue light is again reflected from a silvered mirror into the 'blue' vidicon tube, which generates the 'blue' signal. The red and green light from the blue dichroic mirror is directed towards a red dichroic mirror which reflects the red portion and transmits the green light. The red light is again reflected by a silvered mirror to the 'red' vidicon tube, from which the 'red' signal is obtained. The green light passes through the red dichroic mirror to the 'green' vidicon, and this tube provides the 'green' signal.

* The EIA resolution slide may be obtained from Electronics Industries Assn., 1721 DeSales St., Washington, D.C. A catalogue listing test films and slides is available without charge from the Society of Motion Picture and Television Engineers, 9 East 41st St., New York, N.Y. 10017.

A fixed lens is located in front of each of the vidicon tubes. These lenses are focused on the plane of the camera field lens. In this way, sharply defined images of the film in the projector gate are produced at the face plates of all four vidicon tubes. These images must have exactly the same size, and they must be precisely registered in the vertical and horizontal planes.

Dichroic mirrors usually have spurious transmissions in parts of the spectrum outside the area or band of primary interest. Trimming filters are used to suppress these unwanted transmissions, and limit the light reaching the tubes to relatively narrow bands in the red, green and blue portions of the spectrum.

Some color cameras are fitted with color filters instead of dichroic mirrors, and filters may be selected with any desired passbands, in respect to width and location in the spectrum, consistent with acceptable reproduction of color film pictures. Some designers have adopted the principle of narrow band reproduction, to reduce as much as possible unwanted transmissions in the color film dyes. Others have favored wide-band systems, because telecine cameras of this type are less sensitive to differences in color film dye systems.

Camera Control Unit (CCU)

The signals appearing at the output of the telecine camera must be monitored and processed in a camera control unit before being fed into the television transmission system. Usually the camera control unit is fitted with a small monochrome picture monitor, an oscilloscope or waveform monitor, a group of electrical controls for the camera, and several rows of pushbutton switches for selecting the various modes of operation and displaying the signals on the monitors. At the camera control position a color picture monitor must be provided to enable the operator to adjust the equipment properly and to evaluate the quality of the color pictures being transmitted.

Normally, the three signals from the color camera (four in 4-tube cameras) are displayed side by side on the waveform monitor as shown in Plate 4. It is customary in setting up vidicon cameras to adjust blanking level to coincide with dark current. This level becomes, then, zero signal level, representing a no-light condition, or absolute black. Now, if one of the projectors is turned on, and the light from the lamp is allowed to pass through the picture aperture into the television camera, signal levels are obtained from each of the camera tubes. The levels appearing on the waveform monitor depend on the amount of light entering the camera and the settings of the electrical controls. With the master gain control in its mid-position, the target and beam controls should be adjusted to obtain a signal level of 100 units on the waveform monitor from each of the camera tubes.

With a tungsten lamp in the projector, the amount of energy in the red portion of the spectrum is much greater than that in the blue region. Setting up the camera in the manner just described compensates for this condition. With equal signal levels from the camera tubes, the camera may be said to be adjusted to a neutral condition, irrespective of the spectral emission of the projector light sources. This is known as the open-gate procedure for setting up color telecines.

When a color film or slide is placed in one of the projectors, the signal levels from the camera are reduced considerably, because the amount of light transmitted by the lightest area of the film is always much less than the light reaching the camera through the open picture gate. A typical value of minimum density* in color films is about 0.30. This reduces the light reaching the camera from the projector lamp by 50 per cent.

A convenient way of achieving this condition during telecine alignment is to place a 0.30 neutral density filter in the projector light beam. Then the target and beam controls of the camera can be adjusted for an average film reproducing condition. Neutral density filters can be obtained from photographic supply stores.

Film Variability Problem

The amount of variability, in maximum and minimum densities and in color balance, that can be tolerated in direct screen projection of film is much too great for television. When a film is reproduced, video signal levels rise and fall as the picture densities vary and as the color balance shifts, and these changes in the video signals affect the appearance of the pictures as seen on home receivers.

A long-standing principle in broadcasting is that corrections for abnormal signal levels should be effected at the source. Applied to film telecasting, this would imply that the characteristics of the film images should be altered in such a way that acceptable video signals may be generated with predetermined settings of the controls in the film reproducing equipment. Most broadcasters have come to the conclusion that this is not feasible—suppliers of films seem to be unwilling or unable to make films with a sufficiently small amount of variability for television, or to remake films that obviously are unsuitable for this purpose. To maintain a reasonable degree of uniformity in the video signals broadcast to the public, telecines are provided with controls to compensate for these variations.

Telecine Video Control

Considerable variations in signal levels due to non-uniform minimum densities in films and slides are typical of telecine operations everywhere. Some television stations assign a video operator to the telecine camera control unit to compensate for these variations, and maintain peak signal levels as close as possible to 100 units on the waveform monitor. As the signals from the camera rise or fall, the video operator adjusts the master gain control. This control rotates a neutral density disc in the projector light beam, increasing or decreasing the light entering the camera.

* The term 'density' is defined as the common logarithm of the ratio of the incident to the transmitted light.

Usually, the maximum density in films also varies from scene to scene and from one film to another, over a fairly wide range, and this causes the black level in the video signals to shift relative to zero on the waveform monitor. Another control is available by which black signal level representing the darkest picture areas can be altered to compensate for these density variations.

Equipment has been developed for maintaining peak signal level automatically. In some telecines a separate black level control is also provided, automatically maintaining minimum signal level at a pre-determined value. Facilities of this kind are especially attractive in the smaller television stations where significant savings can be effected by reassigning telecine video operators to other duties. A favorite argument for automatic signal level control in telecine is that this equipment will act more quickly than an operator, and that better uniformity can be maintained, compared with manual operation.

These claims are quite true. In no other area of television operation, however, would this argument be used. Any station adopting automatic signal level control is demonstrating that film is considered to be a second-class programming medium, and that no attempt is being made to improve film performance.

Even in those stations practicing manual control of telecine signal levels, there is often a tendency to try to maintain peak signal levels at all times at 100 units on the waveform monitor, irrespective of the characteristics of the pictures. This is a practice unique in television operations. While there certainly is a need to watch maximum signal levels carefully, to ensure that excessively high levels are not transmitted for any appreciable length of time, it is not necessary, by any means, to raise the signal levels every time a darker scene is encountered, especially when it should be obvious from the appearance of the pictures or the nature of the program, that this is a normal condition.

The practice of 'riding' video signal levels in telecine is the unfortunate outcome of the inability of the film industry to produce television films that will generate uniform signal levels. It is impossible for a telecine video operator to determine whether signal levels are low because the film producer wanted the pictures to be dark, or the low levels are simply the result of carelessness in making the film, or misuse of the film process.

Color Balance Control

Extreme variations in color balance, film-to-film, and scene-to-scene, are typical of 16 mm films especially. To compensate for these variations, it is customary to provide color trim controls in the camera control unit. With these controls, peak signal levels from the red, green and blue camera tubes may be raised or lowered. These adjustments alter the mixture of the three color signals entering the following encoder equipment, and in turn, the modification of the encoded signal output causes the pictures viewed on a color monitor to take on a different appearance. In some color telecines, separate trim controls are provided for the minimum signal levels, to compensate for colored shadows.

The ability to alter the balance of the color pictures in telecine is one of the major advantages of television film programming. Unfortunately, color correction in telecine is used mainly to compensate for errors in exposing the films—errors that should have been corrected in the film laboratory, or at the camera where the original film was being exposed. It is well known that best results are obtained always when the film is correctly exposed in the first place, rather than by attempting to compensate for errors during the reproduction of the film in telecine.

Color trim controls are used to alter the relative amplitudes of the red, green and blue signals from the camera. If the pictures have an excessively

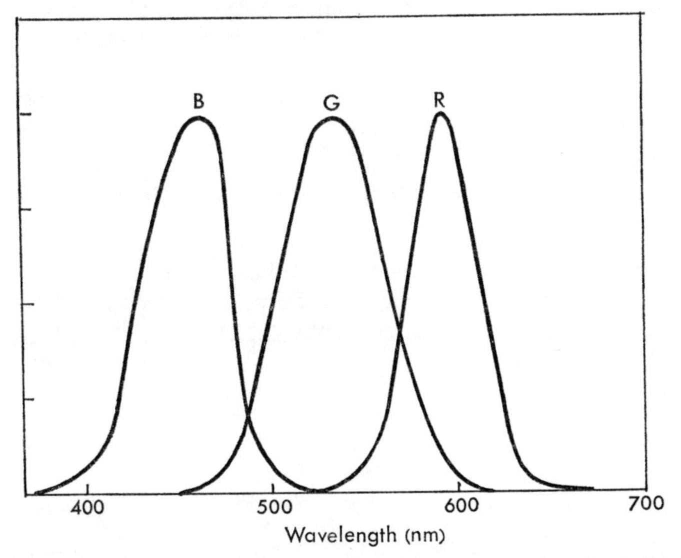

Fig. 25 Spectral transmittance curves of the color separating filters in a typical telecine camera.

reddish appearance, for example, this unpleasant condition can be modified in a more favorable direction by either reducing the gain in the red channel, or alternatively, by increasing the gains in the green and blue channels.

Visual evaluation of color pictures on a monitor may be misleading at times. It may be found, for example, that the reddish appearance of the pictures is due, not to an excess of red, but rather to incorrect exposure of the green layer of the film, resulting in an excess of magenta dye. If that is the case, adjusting the gain of the red channel will not correct the fault; what is needed is more green in the picture, or alternatively, less red and blue.

Sometimes color picture appearance cannot be corrected by adjustment of color channel gain controls. Particularly difficult is a condition where the balance of the highlights is different from that of the shadows. This may be due to some fault in the production of the film, or fading of the dyes in old films. Electronic color correction offers the possibility of altering the color of the shadows more or less independently of the highlights. This can be done by raising or lowering minimum signal levels relative to one another, while the peak signal levels corresponding to picture highlights remain unchanged.

All color films should be rehearsed prior to on-air release to enable telecine video operators to select the most favorable settings of color trim controls. Few stations have the staff or facilities to do this, however. A common practice is to locate a scene with faces near the start of the film, and adjust the trim controls for best reproduction. The film is then allowed to run without further adjustments, whether or not the appearance of the picture changes.

Because of the frequent shifts in color balance between scenes and from one film to another, many stations have given up attempting to compensate for these changes, and simply run the films as received from the suppliers. In these stations it is customary to use automatic signal level control to enable telecines to be run unattended.

European Methods of Color Compensation

European broadcasters have take a more sophisticated approach to the color film problem. It has been proposed* that the color separation signals obtained by scanning the film should be passed through suitable non-linear circuits to secure better tracking of the signals. Equipment designated by the term TARIF has been developed by the BBC in England to effect corrections of this nature.

Further work in this direction has indicated the need to compensate additionally for deficiencies in the film dyes. To effect corrections of this type electronic masking is employed. With this technique equal but opposite errors for the errors due to the dyes are deliberately introduced into the signals from the film scanning device.†

In Italy engineers at RAI have developed equipment by which color film corrections may be stored in a memory device, and applied at the appropriate points in the film as the program is being telecast.

Flying Spot Telecine Systems

In European practice, the flying spot scanner is favored for several reasons. First by speeding up film movement only slightly, from 24 to 25 frames per second, scanning can take place at the television frame rate, thus avoiding a number of difficult mechanical and electrical problems. At the same time, continuous film movement may be employed, affording quieter and smoother projector operation. The quality of the television pictures that can be

* *Electronic Compensation for Color Film Processing Errors* by Wood, Sanders and Griffiths, *SMPTE Journal*, Sept. 1965, pp. 755–759.

† *Some Considerations in the Television Broadcasting of Color Film* by C. B. B. Wood, *SMPTE Journal*, April 1969, pp. 256–260.

obtained with this type of equipment, especially from 35 mm film, is outstanding.

The 35 mm format is favored for use in flying spot scanning systems because the signal-to-noise ratio is dependent on the size of the optical aperture. Inevitably, the results with 16 mm film are poorer, by a factor of about 6 db.

When a flying spot scanner is used to generate color signals, the scanning tube phosphor must have a suitable spectral emission characteristic, and the optical system must be modified by the addition of a beam splitter to separate the light transmitted by the film into red, green and blue components. This is effected usually by an arrangement of dichroic mirrors and trimming filters, as shown in Fig. 26.

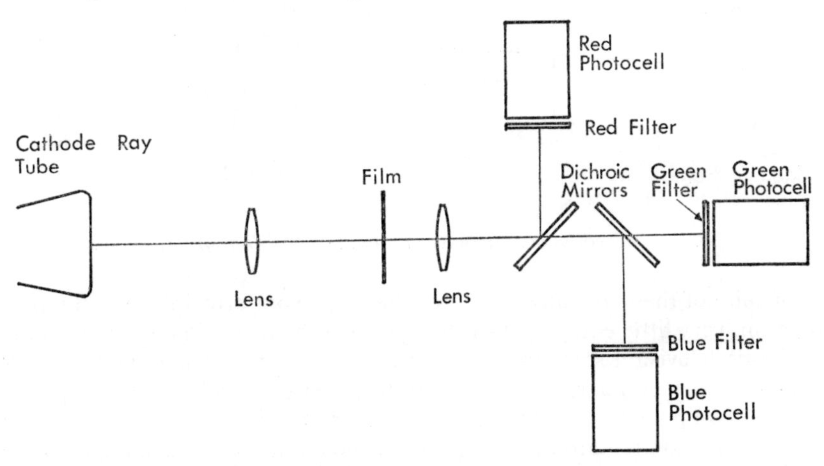

Fig. 26 Essential components of a color flying spot scanner.

An important advantage of the flying spot scanner in reproducing color is that the film is scanned before color separation takes place, thus avoiding color registration errors.

In the twin-lens telecine, which gives especially high quality pictures from 35 mm film, continuous motion of the film is employed, each film frame being scanned by two consecutive television fields, imaged on the film at two different positions in its travel. A split optical system produces spatial displacement of the two fields to achieve interlaced scanning. Movement of the film provides about one-half of the vertical scanning.

Film transport in the flying spot telecine must run in synchronism with the television scanning rate. Because of the strict one-to-one requirement between film frames and television pictures, telecines of this type cannot be used in the North American television system. An alternative method is to make use of a rotating polygonal prism to achieve stationary television pictures from the continuously moving film. However, the unstable optical and mechanical components, as well as relatively low light efficiency in equipment of this type has proved to be an obstacle.

Another technique for utilizing the flying spot principle in 30-frame television systems is that of transporting the film within the vertical blanking interval—that is, in a period of 1.14 milliseconds. By means of a specially designed intermittent mechanism, the film is allowed to remain stationary in the projector gate for alternate periods of two and three television fields. Of the various methods that may be employed to move the film in this short interval, a pneumatic device allows the acceleration forces to be distributed over the entire surface of the film instead of the edges of the perforations.

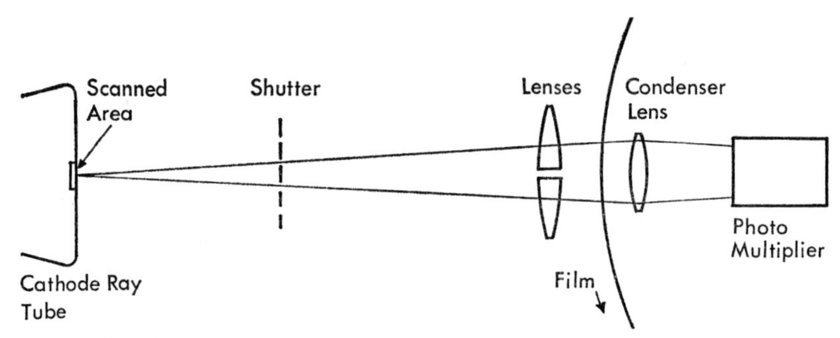

Fig. 27 Essential components of a twin- ens flying-spot telecine.

In spite of the many advantages of the flying spot principle in film reproduction, very little equipment of this type is to be found in countries on the 30-frame television standard. Camera tube telecines—vidicon and Plumbicon—are very popular, due mainly to the lower cost and relative simplicity of the equipment, as well as freedom from film-frame/television-field synchronization requirements. One of the most important advantages of the vidicon telecine is its ability to operate independently of the television scanning system. This permits switching, mixing, supering, and various other effects to be performed without interruption or disturbances in the transmitted pictures.

Film Width Controversy

In the large professional motion picture producing organizations, 35 mm is the standard film width. The major television network centres in the United States have installed 35 mm projection equipment in their telecines, to be able to take prints made directly from the original 35 mm color negatives.

Outside of the network centers, 16 mm is the standard television film size. In spite of the significant improvements that have been made in 16 mm systems in the past 10 or 15 years, this film size has yet to be accepted for serious professional use in the motion picture industry. Only rarely is 16 mm given the same care and attention as the larger professional film size, with the result that most of the 16 mm color films in circulation, and available to television stations for use in programming, suffer from some form of color distortion and degradation.

It is not difficult to demonstrate that 16 mm color film, exposed and processed properly for television use, is capable of a level of performance at least equal to video tape. This level of performance can be achieved easily with original reversal color materials, and with reduction prints from 35 mm color originals. Most of the degradations and distortions encountered in 16 mm films supplied for television use are the result of errors in intermediate duplicating stages.

Recently there has been considerable interest in 8 mm film, especially the Super-8 format. Some television stations have been using 8 mm cameras for news; others have been experimenting with 8 mm for educational programs. The possibilities of automated telecine operation with 8 mm film cassettes are so attractive that this film size will soon come into general use in television, and may eventually displace 16 mm altogether. Systems have already been described* for automatically printing, processing and packaging 8 mm films, with capacities far beyond anything available for the larger film formats. These systems have the advantage that the films are untouched by human hands; perhaps even more important, large numbers of identical prints can be made at very moderate cost, each print having the highest possible quality of which the process is capable.

European Situation

In Europe an entirely different situation exists. There are very few small individual television stations equipped only with 16 mm telecines. European broadcasters rely primarily on 35 mm for the film portions of their program schedules. For events in the news, sports and public affairs categories, however, 16 mm is preferred, mainly due to the smaller size and light weight of the camera equipment, and to a certain extent, the greater ease of processing.

In spite of the obvious quality advantages of 35 mm programs reproduced on flying spot telecines, the successful use of the smaller film format by North American broadcasters represents a challenge that is certain to influence the future use of film in European television. At the same time, camera-type telecines offer the important operational advantage that there is no need for synchronization of film movement with the television scanning rate.

* See *Design Considerations for a High-Efficiency Contact Motion Picture Printer with Magnetic Sound Transfer and Monitoring*, *SMPTE Journal*, Sept. 1967, pp. 904–907.

4 Color film systems

One of the advantages of film as a television programming medium is its flexibility and versatility. In most cases, several different methods may be employed to achieve a given result. Beginning with original camera films, many different types of stock are available, each with its own special characteristics. From the original camera film, several different courses may be taken to obtain the finished prints, ready for use.

By proper attention to the requirements of the different reproducing conditions, excellent color pictures can be produced in the television system, or alternatively, on huge theater screens. The techniques for introducing slow motion, rapid action, animation, optical effects, titles, microscopic images, and many other interesting and unusual sequences are relatively simple and straightforward.

With the aid of special intermediate duplicating materials, 35 mm camera originals may be reduced to 16 mm or 8 mm, and any desired number of copies (prints) may be made from these intermediates while the valuable originals remain safely stored in a vault.

Negative–Positive System

The practice of using a low-contrast negative material for 35 mm professional motion picture production was well established in the earlier black-and-white era, before color film came into general use. To obtain prints from these negatives, with sufficient contrast for viewing on a projection screen, a relatively high-contrast positive material was employed. The main advantage of this practice is the much greater latitude afforded in the exposure of the camera negative as shown in Fig. 28.

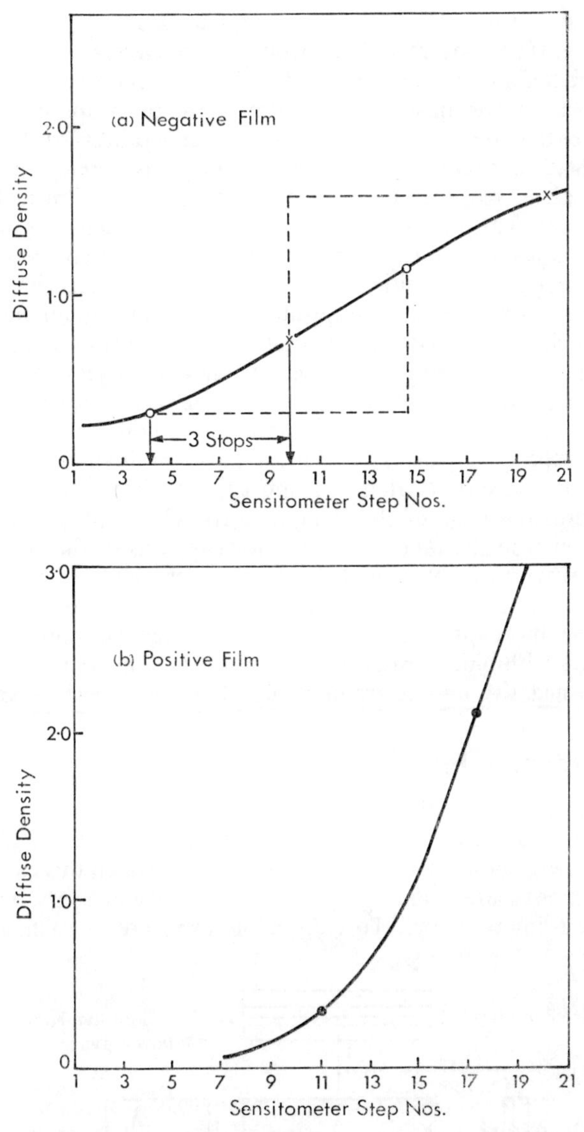

Fig. 28 (a) Black-and-white negative characteristic curve showing maximum and minimum densities of a scene at two exposure levels (b) characteristic curve for a black-and-white positive print film.

The characteristic curve for a black-and-white negative camera film is shown at *a* in Fig. 28. Plotted on this curve are the maximum and minimum densities for a typical scene, at two different levels of exposure, three camera lens stops apart. It can be seen from this illustration that the greater of the two exposures merely increases all of the densities in the negative, without affecting the picture gray scale to any great extent. If the curve had a steeper slope, the exposure latitude would be much less.

The characteristic curve in Fig. 28(b) is representative of a commonly used black-and-white positive print film. Here the slope of the curve is much steeper than the negative in *a*, so that when a print is made from a low-contrast negative, the contrast of the picture images is increased considerably. It is much easier to control the exposure of the print film in the laboratory, compared with the control of exposure of the negative in a camera. By suitably adjusting the exposure of the print film, to compensate for the differences in negative densities, as shown in Fig. 28, prints can be obtained that are very nearly identical in appearance, with maximum and minimum densities as shown plotted on the positive curve. However, the graininess of the print from the over-exposed negative may be objectionable.

A black-and-white negative-positive system is required only to give visually acceptable gray-scale reproductions of colored objects and scenes. To do this, the negative material must have sensitivity to all the colors throughout the spectrum—that is, it must be panchromatic.

The natural response of the light-sensitive silver halide crystals in the film emulsion is in the blue and ultra-violet regions of the spectrum only, but the sensitivity can be extended into the green and red regions as well by the addition of chemical substances to the emulsion during manufacture. For most purposes, panchromatic negative films are needed. Positive films with blue sensitivity only are used normally in making prints, because the negative images are made up of colorless black silver grains.

Color Negative Film

All color processes resolve the various colors in an original object or scene into three separate records, representing the red, green and blue components of these colors. The only practical way in which this can be done is to provide three separate photographic films or emulsion layers, one to record each of the primary colors. To restrict the exposure of each of the three

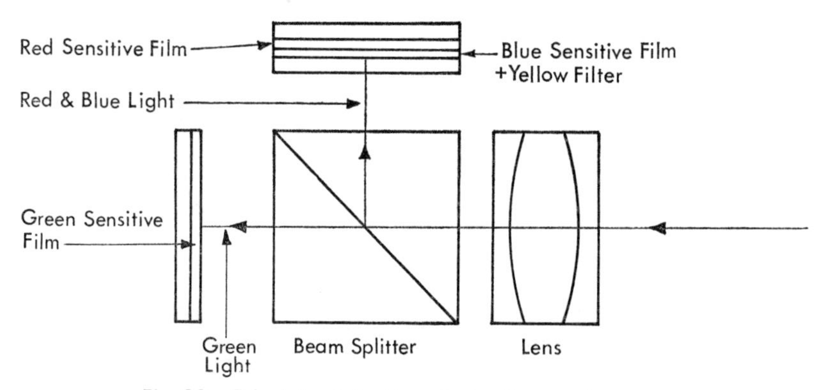

Fig. 29 Principle of the Technicolor three-strip camera.

films or layers to only a single primary color, its sensitivity must be confined to the proper region of the spectrum.

The original Technicolor process made use of three separate films exposed simultaneously in a special camera as shown in Fig. 29. The light entering the camera through the lens is split in a light-dividing prism into two parts. One, consisting of all the green light, passes through the prism to a negative film with high sensitivity to green light. The other part, consisting of the remaining red and blue light, is reflected to an aperture at the side of the camera. In this aperture, two films travel together, emulsion-to-emulsion. One of these records the blue light, while the other is affected only by the remaining red light.

After development, these three films contain black silver images representing the amounts of red, green and blue light in the original object or

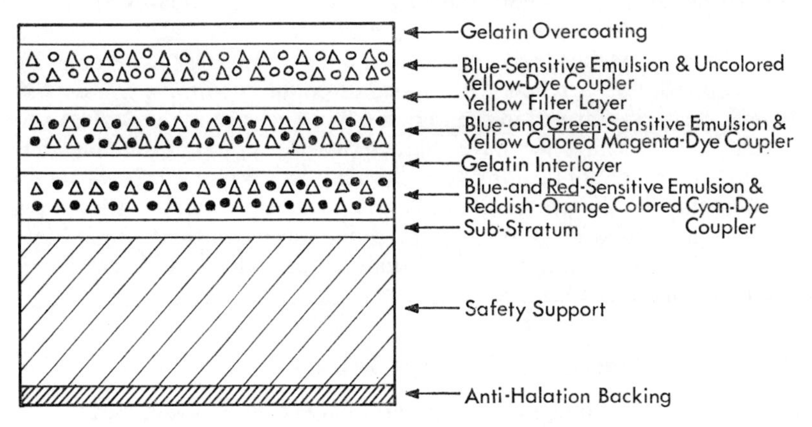

Fig. 30 Schematic diagram of a multilayer color negative film.

scene. From these negatives, color prints may be made in any one of several different ways, one of which is the famous Technicolor process, in which the images are formed by a dye imbibition method.

By coating the red, green and blue emulsions as super-imposed layers on a single film base, the film may be exposed in a conventional motion picture camera. When the process is so arranged that these three layers yield negative *color* images, the film is known as multilayer color negative. The color images of a multilayer color film are composed of dyes instead of silver. This is necessary in order that the characteristics of each of the three layers may be controlled independently of the other two. When the three layers are coated on a single support, and cannot be physically separated from one another, color dye images allow the information contained in each layer to be translated to the color positive film in such a manner as to produce the proper amounts of dye in the print images.

The structure of a typical multilayer color negative film is shown diagrammatically in Fig. 30. The top layer is blue-sensitive only, and records all of the blue light entering the camera from a colored object or scene.

Beneath this layer a yellow filter layer is coated, to prevent blue light reaching the remaining two layers. The middle layer is blue and green sensitive, but because blue light cannot reach this layer, it records only the green light. The bottom layer, sensitive to blue and red light, records only the red light.

Most multilayer color films have substances known as color couplers uniformly distributed in the three emulsion layers.* A different type of color coupler is incorporated in each layer. As the film is processed, the color couplers are converted by chemical action into dye images. In the upper blue-sensitive later a yellow dye is formed; in the green-sensitive layer a magenta dye, and in the bottom red-sensitive layer, a cyan dye. These are the complementary colors of the original red, green and blue primaries.

In the processing of a color negative, silver and dye images are formed during a single developing step. As the exposed silver halide crystals are being converted to black silver grains, the developing agent becomes partially oxidized, and it reacts with the color couplers to form dye deposits. These deposits are formed in the region of the silver halide crystals, so that the resulting dye image will have characteristics as nearly as possible identical with the original optical image by which it was produced. Subsequently, at a later stage of the processing cycle, the black silver images are removed by a bleaching solution, leaving colored dye images in their place. Finally all of the remaining unexposed and undeveloped silver halide crystals are removed by a fixing solution.

Characteristics of Color Dye Images

In films made up of dye images, the subtractive colors—cyan, magenta and yellow—are used to control the red, green and blue components of the light used in printing or projection. The amount of cyan dye determines the amount of red light passing through the film, since cyan subtracts red from white light, while at the same time green and blue are transmitted. Similarly, the amount of magenta dye controls the green light transmission of the film, since magenta absorbs green and allows red and blue to pass through. Finally, the amount of blue light transmitted by the film is controlled by the yellow dye, which absorbs blue and transmits freely the red and green.

The amounts of each dye remaining in the film after processing are dependent almost entirely on the level of exposure for each layer, and for each small area in the individual layers, representing all of the colors in the original object or scene. A bright red object, for example, would be represented by a heavy deposit of cyan dye in the red-sensitive layer, while at the same time there would be minimal deposits of magenta and yellow dyes in the green and blue sensitive layers. This means that in a color negative, all of the colors of the original object or scene are reversed. In the

* In some types of color film the color couplers are in the processing solutions.

following color print stage, a similar procedure of image reversal takes place, leaving dye images that have an appearance similar to the original object or scene.

The dyes used to form the images in subtractive color films do not have ideal absorption characteristics, and they do not transmit in an ideal manner in the remaining portions of the spectrum. This leads to undesirable results when a color negative is printed on a color positive print material. For example, the cyan dye, which should absorb only red light, also absorbs some of the blue and green.

Spectral density curves for a set of typical color film dyes are shown in

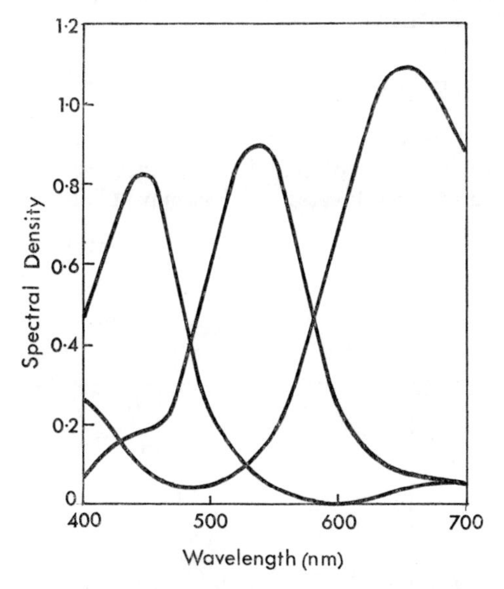

Fig. 31 Spectral density curves for the cyan, magenta and yellow layers in a color film.

Fig. 31. These curves are obtained by measuring the dye images in a spectrophotometer.*

A relatively simple method for eliminating the effects of the overlapping absorptions of the dyes is the use of colored couplers. In the Eastman Color Negative Film, Type 5254, the coupler in the red-sensitive layer is colored orange, while the green-sensitive layer contains a yellow-colored coupler. When the coupler in the red-sensitive layer is converted to cyan dye the orange color is destroyed. The unexposed areas, in which no dye-conversion takes place, retain their orange color. Areas of intermediate exposure contain some cyan dye and residual orange coupler. This results in a

* Methods of measuring color images are described in the book *Principles of Color Sensitometry* published by the Society of Motion Picture and Television Engineers, New York, N.Y.

cyan negative image and in the same layer an orange-colored positive image composed of the residual coupler.

In the green-sensitive layer, the yellow coupler is converted to magenta dye in areas that have been exposed. Here again, two images are formed in the layer—a magenta negative image and a yellow positive image. The presence of the orange and yellow colored couplers in the processed negatives gives the film an overall orange cast. This is eliminated in the printing stage by suitable sensitization of the print film and the proper selection of light intensity in the red, green and blue portions of the spectrum. At the same time, prints made from negatives of this type are free of the color degradations and distortions caused by overlapping absorptions of the cyan and magenta dyes in the negative.

Positive Color Print Film

The structure of a positive print film is basically similar to the previously-described negative film, except that it is not necessary to have the same arrangement of emulsion layers. In Eastman Color Print Film, Type

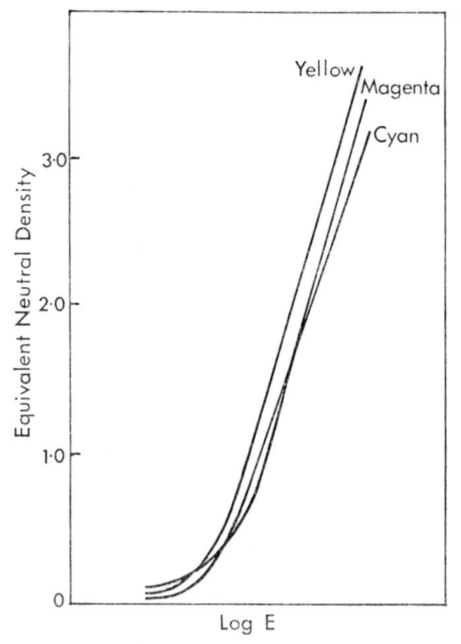

Fig. 32 Characteristic curves for a color print film.

5385, for example, the blue-sensitive layer, containing the yellow coupler, is coated next to the base. This is desirable because the yellow dye image contributes less than the others to picture sharpness, and the bottom layer in a multilayer film is always least sharp. Next is the red-sensitive layer containing the cyan coupler, and at the top is the green-sensitive layer containing the magenta coupler. Also incorporated in these layers are red and green absorbing dyes to reduce the effects of flare. These dyes are destroyed during subsequent processing. The couplers are themselves colorless in order that the finished prints may have a neutral appearance in the white, black and gray areas of the color pictures.

Color Intermediate Films

In the production of motion pictures in color it is often necessary to prepare duplicate negatives from the original camera negatives. Scene-to-scene transitions such as fades, mixes, wipes and many other forms of special effects may be introduced during the preparation of the duplicate negatives. The need for more than one printing master calls for the production of an intermediate negative. It is essential that duplicate negatives made for these purposes have characteristics which will allow prints to be made matching prints from the original negatives.

Improvements in the color characteristics of dye images have made possible a four-stage process consisting of: original color negative, intermediate color positive, intermediate color negative, color print. As a rule the same type of film stock is used for both intermediate positive and intermediate negative. Some films of this type, such as Eastman Color Intermediate Film Type 5253, have colored couplers similar to color negative films, to reduce unwanted dye absorptions, and improve color reproduction. These films normally have medium contrast, to avoid increasing picture contrast in these intermediate duplicating stages.

Film of this type is also available in 16 mm width to obtain reduction duplicate negatives from 35 mm camera originals. These 16 mm negatives are made in an optical reduction printer from a color intermediate positive.

Systems for Producing 16 mm Color Prints

The production of 16 mm motion pictures in color has taken an entirely different course compared with the professional 35 mm sector of the industry. So far there is only limited commercial use of the color negative-positive system in 16 mm film production, although both negative and positive color materials are available in the 16 mm width.

The production of 16 mm color films for educational, training, advertising and industrial purposes was at first a by-product of the introduction of practical color films for home movies in the mid-1930's. These were reversal-type films, giving positive color pictures directly from the original camera film. Semi-professional use of these amateur films soon created a demand for another film material with which color prints could be made, but commercial users discovered that camera films designed for amateur use failed to meet commercial requirements. This led to further development of several families of 16 mm reversal films, capable of yielding the very finest color picture quality.

A system of 16 mm color motion picture production in general use involves a reversal camera original film, and a reversal duplicating material, as well as a reversal color print film. When only a limited number of copies is needed, prints may be made directly from the camera original film. To obtain optical effects, such as fades and dissolves, and supered titles, the A and B roll printing technique may be employed, avoiding an intermediate duplicating stage. High volume release printing requires an intermediate reversal printing master. This system has the advantage that all of the stages involve reversal positives allowing visual evaluation of the color pictures for exposure and color balance.

Reversal Camera Films

The structure of a typical reversal color film is shown in Fig. 33. This structure is similar to color negative materials, and color couplers are

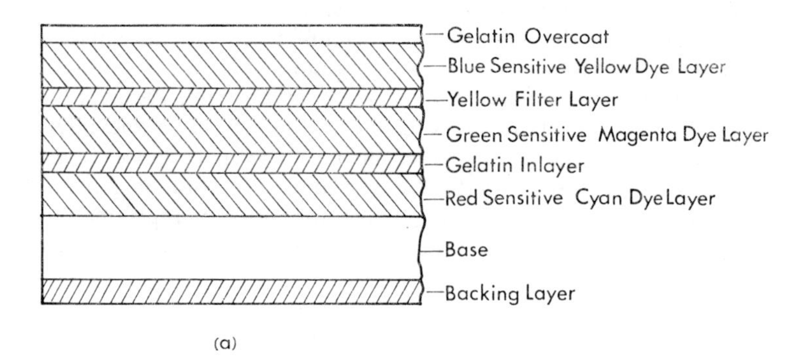

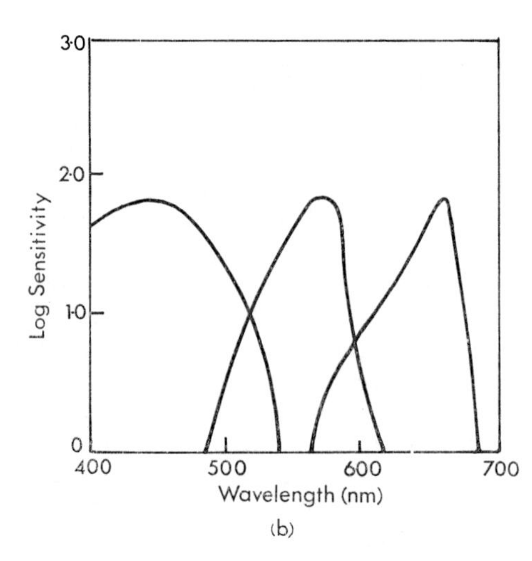

Fig. 33 (a) Structure of a reversal color film (b) spectral sensitivities of the three emulsion layers.

incorporated in each layer. To obtain positive color images in the original camera film, a different method of processing is employed. First, the exposed silver halide crystals in the three layers are converted to black silver grains in a conventional negative developing solution. After a thorough washing, all of the remaining undeveloped silver halide crystals are exposed to light, and the film is then subjected to a color developer. In this stage color dye images are formed. Then the black silver images are removed in a bleaching solution, leaving positive dye images in the film.

The overall sensitivity of color films to exposure is, relatively speaking, quite low. Continuous efforts have been made by the manufacturers to produce original camera films with higher speed. One of the reasons for the demand for high-speed camera films is the practice, fairly general in 16 mm photography, of using available light for the exposure of the film. The trend in the professional 35 mm field has been towards finer grain and better color image quality rather than higher speed, because of the practice in 35 mm film production of using artificial lighting to obtain the desired image characteristics, when available light is insufficient for the exposure of medium-speed materials.

Reversal color films in the 16 mm width are now available with exposure index ratings up to 500 or more. Most of these films can be specially treated in processing to still further increase the speed ratings, although this is likely to affect adversely the quality of the color pictures.

Color Internegative Films

It is generally agreed that the finest quality 16 mm release prints are obtained by printing directly from the 16 mm reversal camera originals on a

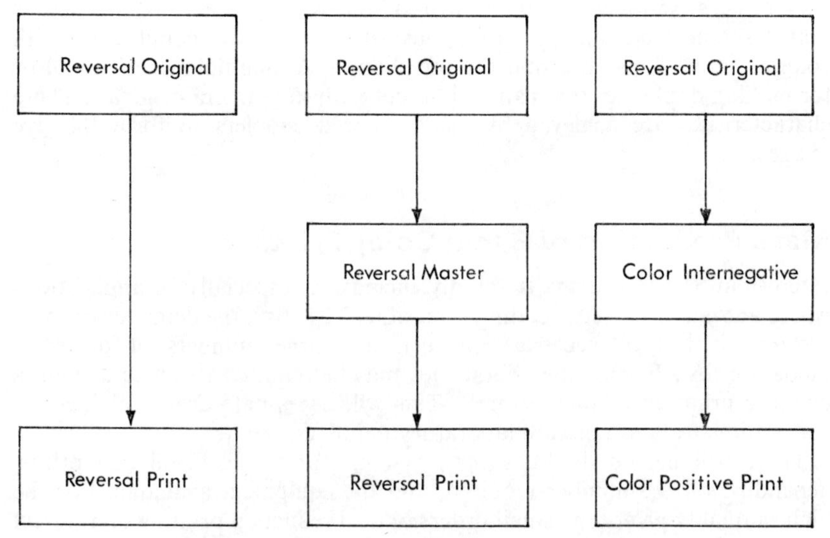

Fig. 34 Duplicating systems for 16 mm color films.

color reversal print film. However, when a large number of release prints is required, this is an expensive process, and there is considerable hazard to the original materials.

Some improvement was achieved by using the reversal color print film as a reversal master, treated to reduce the contrast. Although this method lessened the hazard to the original materials, it did not significantly lower the cost, and considerable degradation of picture quality resulted.

With the aid of a color internegative film, color prints can be made on color positive print film, with much less picture degradation, and reduced cost. Eastman Color Internegative Film, Type 7270, makes use of colored couplers in a manner similar to color negative film to reduce unwanted dye absorptions. In addition the film layers contain soluble dyes which reduce light scattering and thus help provide sharper images. The contrast of this material is quite low, and the graininess is less than that of original camera films. The different courses that may be taken to obtain color prints from reversal color originals are shown in Fig. 34.

Reversal Intermediate Films

Until recently, the use of reversal color materials has been limited to 16 mm film production. In 1968 Eastman Color Reversal Intermediate Film was announced, available in 35 mm (Type 5249) and 16 mm (Type 7249) widths. With this film duplicate negatives of original color negatives can be made in a single printing-processing operation. The elimination of one printing step reduces grain size and improves color reproduction and sharpness. The system is thus particularly useful for obtaining 16 mm reduction prints from 35 mm color originals.

A basic sensitometric requirement of the reversal intermediate film is that it should have unity printing contrast when used in conjunction with color negative films, and that the toe shape and minimum densities allow for making duplicates that can be intercut with the camera originals. These characteristics are achieved by using colored couplers to form the dye images.

Mass Production of 8 mm Color Prints

Interest in 8 mm systems is rapidly increasing, especially in applications where automatic cassette loading is feasible. To satisfy the demands of these systems, it will be necessary to turn out large numbers of prints at moderate cost. Eventually laboratories may be required to make a million or more prints of a single subject. This will necessitate drastic changes in conventional motion picture laboratory practices.

There will be, no doubt, several systems for producing 8 mm prints, depending on the number required, and the equipment available. For the highest quality prints and small orders optical reduction printing on reversal film may be the most practical. For large orders running into thousands, contact-release printing using 35 mm/Super-8 film will be particularly advantageous. A typical method of operation is shown in Fig. 35.

Film is now available 35 mm in width, but with five rows of perforations, which permits four Super-8 prints to be made simultaneously by contact printing from a 35 mm internegative. This internegative is prepared by optical reduction printing from a 16 mm color original. Either optical or magnetic sound tracks may be laid down on the prints during the production operation. After the 35 mm prints have been processed the

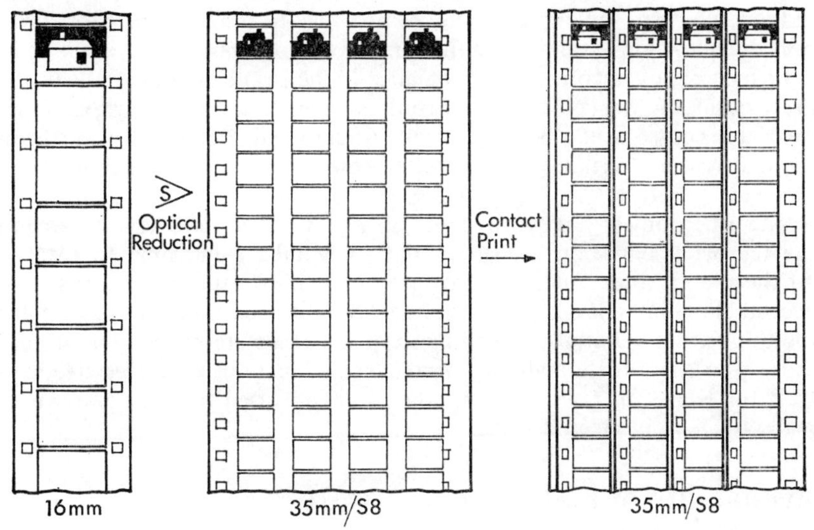

Fig. 35 35 mm/Super-8 printing system and film formats.

film is slit to obtain the individual Super-8 prints. Systems have been proposed capable of turning out Super-8 prints at the rate of 800 ft per minute, in which picture printing, magnetic sound transfer and sound monitoring take place simultaneously. The special 35 mm print film shown in Fig. 35 is magnetic pre-striped.

Color Film Systems for Television

Television broadcasters in all parts of the world make extensive use of film. Some broadcasting organizations, relying mainly on all-electronic systems, supplement live and video tape programs with film, while others, particularly the major networks in the United States, give film first place in prime time programming periods. The quality of the 35 mm film programs released from these network centers is, for the most part, excellent. Conventional motion picture materials are used for all of these productions, which are usually color prints made directly from the 35 mm color camera originals.

Outside of the major US network centres, an entirely different situation prevails. For the most part, the remainder of the television industry makes use of 16 mm color prints obtained through film distributing organizations. The quality of much of this material is very poor indeed.

The major criticisms of currently available color films have been excessive contrast and incorrect color balance. To compensate for incorrect color balance, various proposals have been put forward for electronic color correction methods. All of these require pre-screening of the film on a standardized reproducing system, and the selection by a skilled operator of appropriate electronic correction factors to compensate for the color errors in the film. Then, some means must be provided to program the corrections into the television system while the film is being telecast.

In most cases, incorrect color balance is the result of some fault in the production of the color print. From an economic standpoint, a more favorable course would be to insist on correctly balanced color prints from the distributor. Certainly, much better color pictures can be obtained in this way.

The question of excessive contrast has aroused a lively debate in the television and film industries that may take some time to resolve. At the present time there are two opposing camps—those who insist that acceptable color television pictures can be obtained with available color motion picture materials, providing these materials are properly used; and those who claim these materials produce color pictures with excessive contrast for television reproduction, when they are used to photograph normally lighted objects and scenes.

Availability of materials

The references in this chapter to specific materials are for illustrative purposes only, showing how the characteristics of those materials may be applied to the intended purposes. Other materials with similar characteristics may, of course, be used in a similar manner and some materials may have characteristics enabling different results to be achieved.

5 Lighting and exposure for color

When a multilayer color film is exposed in a camera, the three emulsion layers are affected differently, depending on the amounts of red, green and blue light reaching the film. If, for example, an area in a scene reflects only red light, an image will be formed only in the red-sensitive layer of the film, but the red light will have little or no effect upon the green- and blue-sensitive layers. White light, containing all three colors, will affect all three layers of the film.

Light which appears white to the eye, however, may be quite different in spectral composition. Daylight and tungsten illumination, for example, contain different amounts of red, green and blue. While the eye adapts easily to these two quite different kinds of white light, the color film records the relative amounts of red, green and blue in the white light, in the form of images with different magnitudes. For this reason the manufacturers of color films normally supply two different types—one balanced for average daylight, and the other for average tungsten illumination. Alternatively, filters may be placed over the camera lens to alter the spectral composition of the light entering the camera, and match the spectral sensitivity characteristics of the film in use.

The amount of exposure the film receives must be carefully controlled to record scene elements on the film properly. When an outdoor scene is photographed, light from the sun and sky is reflected in different ways by the various elements making up the scene. Some parts of the scene may be very bright, such as clouds, snow or white buildings. Other areas may be dark, reflecting very little light to the camera, or these areas may be shaded from the direct rays of the sun.

Every scene contains what might be termed a scale of grays, from white objects to dark shadows, with various intermediate shades between these limits. With a color film in the camera balanced for daylight illumination, this scale of grays produces images in all three emulsion layers. In addition, most scenes contain colored objects which will produce images that predominate in one layer, or, if the colors are highly saturated, the images will be confined to a single layer. Most objects reflect light of all colors, the color that predominates giving the object its characteristic appearance. Green grass, for example, is not 'pure' green. If the layers of an exposed and processed color film could be separated it is likely that images of grassy areas in a scene would be found in all three layers, with the image in the green sensitive layer predominating.

Film Speed

The sensitivity of a photographic material to exposure is known as its speed. Film speed is expressed as a number related to the response of the material. Many different methods of speed rating have been employed,

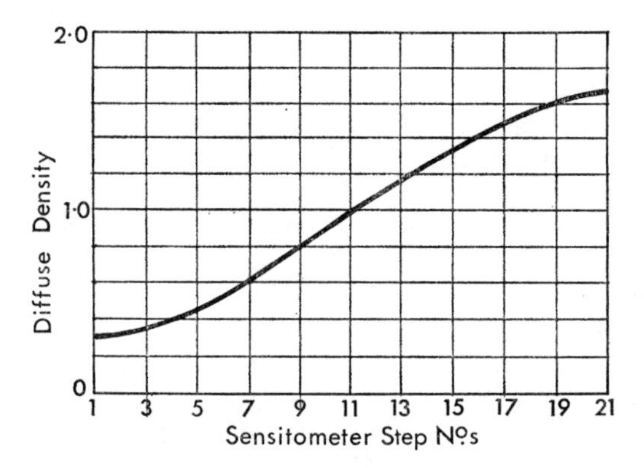

Fig. 36 Characteristic curve for a black-and-white negative film.

and still there is not complete agreement on a universal speed rating system for all materials and all applications.

The main purpose of a speed rating is for use with photo-electric exposure meters. With the aid of a meter a number may be derived indicating the amount of light falling on a scene or being reflected from it. This number, together with the speed rating from the film, may then be used to set the aperture of the camera lens so that a suitable amount of light will reach the film. The need for such a system is evident when it is considered that the light illuminating a scene may vary over an extremely wide range, and there are a great many different types of film with widely varying sensitivities.

The response of the film to exposure may be illustrated graphically, in an easily understandable form, by means of the characteristic curve. To obtain the characteristic curve, the film is placed in a device known as a sensitometer, and given a series of exposures increasing by a factor of $\sqrt{2}$—that is, 1.414—for each step. After the film has been processed, the densities of the steps are measured with a densitometer and plotted on graph paper. By connecting the plotted points with a smooth line, the characteristic curve is produced, as shown in Fig. 36. This curve represents the response to exposure of a typical black-and-white negative film.

The concept of film speed has obvious psychophysical connotations. One film is regarded as having a higher speed than another when it will produce satisfactory photographic images with less exposure. An American Standard method for determining photographic speed and exposure index, Z38.2.1, was adopted in 1947, specifying that 'the speed of a negative film for continuous-tone black-and-white pictorial photography is inversely proportional to the minimum exposure required to produce a negative from which a positive print can be obtained that will give an image of high quality'. As such it was a subjective method, but sensitometric test strips developed with the picture negatives were used to establish the conditions under which acceptable pictures could be obtained.

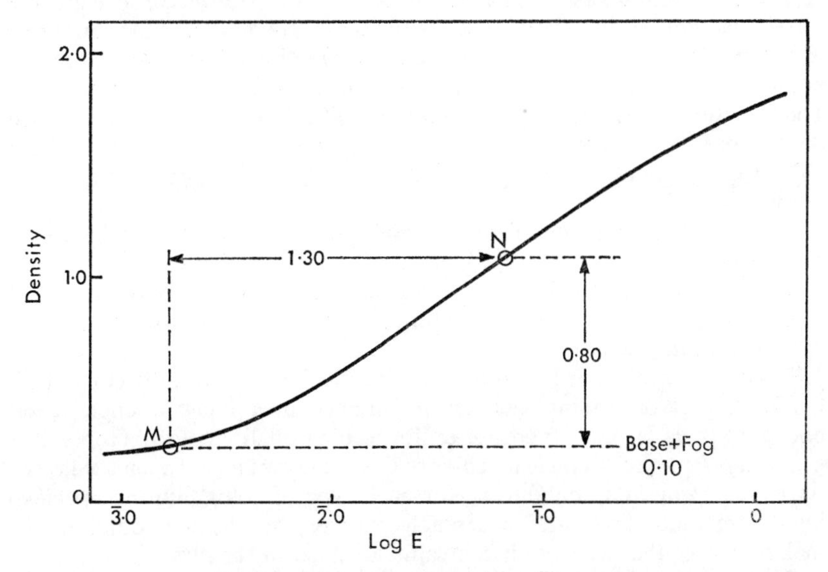

Fig. 37 Method for determining film speed.

This standard was based on what was known as the fractional gradient criterion—that is, best pictures were obtained when the darkest scene areas were reproduced in the toe of the characteristic curve at a point where the gradient was 0.3 of the average gradient over a log exposure range of 1.5.

A new standard, adopted in 1960, makes use of a much simpler concept—namely, that speed shall be determined from a point in the toe of the characteristic curve where the density is 0.10 above base and fog density. American Standard method for determining speed of photographic negative materials

(PH2.5—1960) gives the following definition of standard film speed, S_x, in terms of the sensitometric test exposure—

$$S_x = \frac{0.8}{E_m}$$

where E_m is the exposure in meter-candle-seconds corresponding to point M in Fig. 37.

In 1961 the corresponding German standard DIN 4512 blatt 1, was revised, and in 1962 the British Standards Institution adopted a similar speed rating method, representing general agreement on the 0.10 density-above-fog criterion.

Exposure Index

These standards are intended for use only with negative films for black-and-white pictorial photography—motion picture films are excluded. However, cinematographers everywhere make use of photo-electric exposure meters calibrated in accordance with these standards. Manufacturers of motion picture films have adopted the practice of assigning numbers known as exposure index values to materials intended for use in cameras. The exposure index values given in film specification sheets are determined on the basis of practical picture tests, but users are warned that these index values should not be regarded as numbers which express the absolute sensitivity of the film.

In view of the different formulas and processing conditions in use in black-and-white motion picture laboratories, the effective speed of a particular film may be quite different. As a result, cameramen have been obliged to modify the values given by the film manufacturers to suit the conditions encountered in practice.

Processing conditions for color films are much more carefully controlled. Uniformity of processing must be maintained in not just a single layer, but in all three layers. To ensure uniform, high-quality color pictures, it is necessary to preserve a delicate balance in each layer and between the layers. As a rule, processing conditions for each type of color film are specified by the manufacturer, and acceptable pictures are usually obtained by making use of the exposure index value assigned to the film.

Data sheets supplied with color films suggest that the exposure index values may be used with exposure meters calibrated in accordance with the appropriate standards. These values also will apply if a meter reading is taken from a gray card with about 18 per cent reflectance, held close to and in front of the subject, facing the camera. For *exceptionally* light or dark colored subjects, the exposure should be decreased or increased, respectively, from the camera lens setting indicated on the meter.

In December 1964 a proposal was published in the Journal of the Society of Motion Picture and Television Engineers that the American Standard method for determining speed of reversal color film for still photography

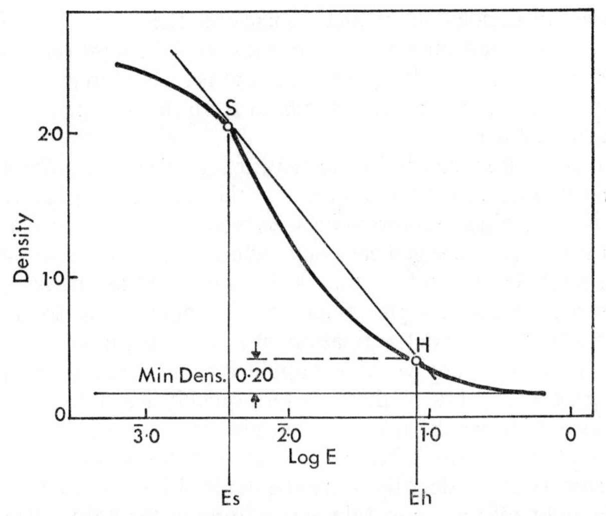

Fig. 38 Method for determining speed of reversal color films.

(PH2.21—1961) should be adopted for reversal color motion picture films as well. This method is illustrated in Fig. 38.

Point H is located on the curve at a density of 0.20 above the minimum density. From Point H a straight line is drawn tangent to the curve at Point S. The exposure E_h and E_s, are used to compute the exposure E_m by the following formula:

$$E_m = \sqrt{E_s \cdot E_h}$$

The exposure E_m represents the sensitometric parameter from which speed, S_x, is computed:

$$S_x = \frac{8}{E_m}$$

Exposure Meters

The simplest form of photo-electric exposure meter consists of a barrier-layer cell connected to a microammeter, with an exposure calculator in the form of multiple rotatable discs. These cells are made from small rectangular iron plates coated with selenium or some other substance with sensitivity to light. When light falls on the sensitive surface an electrical potential is created between the two layers of the cell, and current will flow in the external circuit, consisting of the winding in the meter coil. The microammeter scale may be calibrated in any convenient series of numbers.

Many different types of exposure meters have been designed and placed on the market, but most of these may be divided into two main groups—meters for measuring reflected light, and meters for measuring incident light. In use, an incident light meter measures the amount of light falling on a scene, while a reflected light meter measures the light reflected towards the meter from the scene. (If a source of light is located in a scene, or within the acceptance angle of the meter, the light from this source will be included in the measurement.)

When an exposure meter is directed towards a scene, the different amounts of light reflected by the various objects in the scene are integrated, and the meter reading obtained represents a weighted average of the illumination. The weighting can be changed easily by including more or less sky within the acceptance angle of the meter—that is, by simply tilting the meter upwards and downwards, while the movement of the meter indicator is observed. This method of exposure determination, although obviously unsound from a theoretical viewpoint, enjoys wide popularity and gives surprisingly good results in practice. Users of these meters quickly learn to avoid the misleading effects of large sky areas and bright light sources on the readings. Reflected light meters work best with average outdoor scenes.

Measurements of incident light may be made either by directing the meter towards the light source, or by taking a reading of the light reflected from a white or gray card placed to receive maximum incident light and applying an appropriate factor for the reflectance of the card. The first method requires the use of a special type of meter with a diffusing medium over the cell. Some types of meters are fitted with a translucent hemisphere as a light collector.

In the use of an incident light meter it is assumed that most scenes contain a variety of objects and surfaces reflecting light in a more or less average manner. The simplest case is an outdoor scene with the sun behind the camera.

When artificial light is used to illuminate a scene the exposure situation becomes much more complex, and it is for this type of work that incident light meters have their greatest value. With a meter of this type, key and fill light may be properly balanced, and the overall level of illumination may be adjusted to suit the film in the camera.

Exposure of Color Films for Television

The objective in all methods of exposure calculation is to locate the scene brightness (luminance)* range approximately in the central part of the film's characteristic curve. In practice there is a good deal of latitude in the calculation of exposure, mainly due to the tolerance of the eye in accepting pictures with some black or white compression—that is, loss of detail in shadows or highlights—when the film is under- or over-exposed.

The characteristic curve shown in Fig. 39 is representative of a typical reversal color film. Plotted on this curve are the maximum and minimum densities for a typical outdoor scene, properly exposed on this color

* Brightness is a sensation produced in the eye by light, while luminance is a physical quantity capable of measurement.

film. The effects of increasing or decreasing the exposure by one camera lens stop are shown also. When the exposure is increased, the effect is to shift the scene downwards on the curve, with the high-lights being recorded in the toe where there is little density separation between scene elements. Alternatively, if the exposure is decreased, the effect is to shift the scene upwards, into the shoulder of the curve where shadow details are lost.

Exposure latitude is somewhat greater for negative films, because they have lower contrast—that is, the slope of the characteristic curve is not as steep. Moreover, adjustments can usually be made in printing to compensate

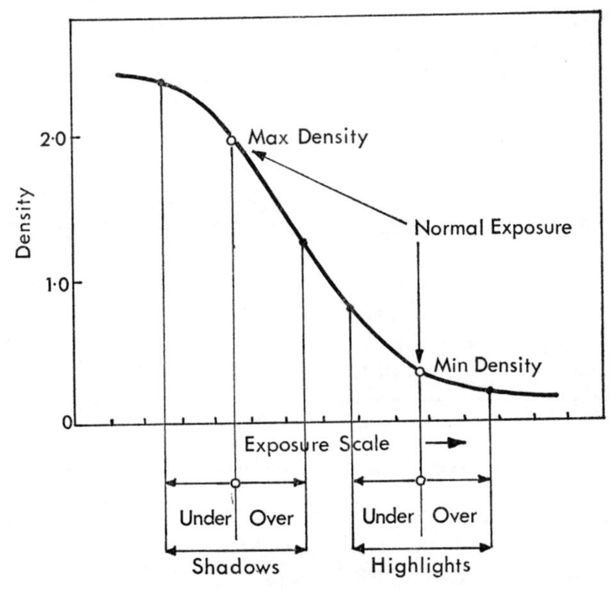

Fig. 39 Characteristic curve for a reversal color film, showing effects of over- and under-exposure.

for under- or over-exposure of negatives. These adjustments make the pictures lighter or darker, but cannot compensate for black or white compression in the picture tonal scale due to incorrect exposure of the negative.

An exposure meter averages the light reflected from different parts of the scene, and may give a similar reading for a bright sunlight scene, and the same scene on an overcast day. While in both cases the average reflectance is the same, the location of the lightest and darkest parts of the scene on the film's characteristic curve are different. This is due to the difference in the lighting contrast. Similarly, a scene with an average reflectance of, say, 12 per cent may have large areas of very high brightness—sand, water, snow or sky—and at the same time large dark-colored areas reflecting very little light. Again, the same meter reading may be obtained from a scene made up of only moderately light and dark areas, and no extremes. Here, too, the location of the lightest and darkest areas in these scenes on the film's characteristic curve is different. This time, the difference is due to subject contrast.

In normal motion picture production, picture areas such as faces and other objects of primary interest should be projected on the screen at a fairly uniform brightness level (unless special effects are desired). The eye readily accepts fairly large variations in the maximum and minimum densities of the picture images, because it has a tendency to compensate for both brightness level and brightness differences.

When a film made in this way is projected into a television camera, signal output level is related to the amount of light reaching the camera tube. Thus, areas in the film transmitting the greatest amount of light (picture highlights) produce the highest signal levels. Peak signal levels cannot be allowed to exceed 100 units on the video waveform monitor scale (see p. 37) but picture areas representing very bright objects or very brightly lighted scene elements may give signal levels rising far beyond the peak white limit. When this happens, the amount of light reaching the camera or the gain of the signal generating system must be reduced, with the result that faces and other picture areas of primary interest appear much too dark.

At the other end of the picture tonal scale, zero signal level is established by cutting off all light from the camera. When a film with varying maximum densities is projected into the camera, black level in the output signals rises and falls as the amount of light reaching the camera in the dark scene areas changes.

Controlling Maximum and Minimum Densities

By far the most serious problem in the use of film in television is the variability of the signal levels from telecine, scene to scene, and film to film. To compensate for these variations, an operator must be assigned to telecine to adjust the controls manually, or alternatively, an automatic signal level control device may be used. With either method, some degradation in the quality of the pictures is inevitable.

It would be highly desirable to reduce these variations to the point where films could be telecast without constant attention by an operator or the use of automatic signal level control. Then film programs could be broadcast in much the same manner as programs recorded on video tape. Of course, at the same time the degradation of picture quality caused by varying the signal levels could be eliminated.

When television programs are recorded on video tape, the signals generated in the cameras are displayed on a waveform monitor, and lighting or camera exposure is modified as required to obtain uniform peak white and black levels as shown in Plate 1. As the signal levels are adjusted to coincide approximately with the upper and lower limits of the waveform scale, the effects of these adjustments on the appearance of the pictures may be evaluated by observing a picture monitor. In this way best possible picture appearance can be achieved, while at the same time uniform maximum and minimum signal levels are maintained. Consequently, there is no need for either manual or automatic adjustment of signal levels when the video tape program is broadcast.

The key to uniformity of video signal levels from film in telecine is proper control of maximum and minimum exposure levels at the time of exposure of the original film in the camera. If the lightest and darkest areas of scenes could be reproduced always at the same points on the film's characteristic curve, the original film—or prints made from it—would generate video signals with uniform peak white and black levels. To achieve this highly desirable result, the film cameraman must be provided with some means for locating and measuring the lightest and darkest areas in scenes, and for suitably adjusting the exposure to reproduce these areas at the specified points on the characteristic curve.

Spot Brightness (Luminance) Meters

The easiest way to locate the lightest and darkest areas of a scene is to make use of a small television camera looking into the viewfinder of the film camera. A combined system of this kind could be calibrated to show maximum and minimum signal levels on the television cameras's waveform

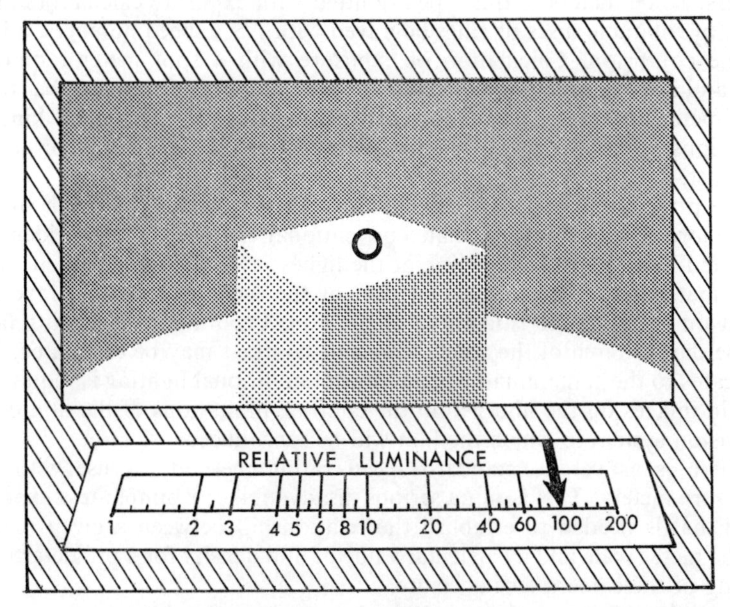

Fig. 40 Use of a spot photometer for calculating exposure. Circle indicates area being measured.

monitor when the film is properly exposed—that is, when the lightest and darkest areas of the scene are reproduced at the specified points on the film's characteristic curve. Such systems are technically feasible, incorporating small vidicon or Plumbicon tubes. Film cameras constructed on this principle are commercially available.*

An alternative method of television film exposure control makes use of a device for measuring the brightness (luminance) of small scene areas. In recent years a number of different types of instruments known as spot photometers or spot brightness meters have been developed for this purpose. These instruments are somewhat more expensive than conventional photoelectric exposure meters, but the advantages of the spot meter in the television film exposure application far outweigh the additional cost. Moreover, a spot meter is much easier to use than an ordinary exposure meter, especially in setting up artificial lighting conditions.

The field of view of a spot meter with an acceptance angle of one degree is shown diagrammatically in Fig. 40. With a meter of this type, measurements can be made in small scene areas such as a person's forehead at a distance of 10 or 15 ft.

Calculating Exposure with a Spot Meter

Film cameramen, accustomed to the use of incident or reflected light meters for determining exposure, are likely to be confused and frustrated by spot meters. Some meters of this type are fitted with exposure calculators in the form of rotatable discs or rings engraved with ASA speed numbers, adding to the confusion. Calculation of exposure with a spot meter requires a new and entirely different approach. It is very important to remember that a spot meter provides a means for correctly exposing the film in all kinds of situations—even the most difficult—irrespective of the lighting or subject contrast.

The principle of exposure calculation with a spot meter is much simpler and more straightforward than conventional methods with incident- or reflected-light meters. A reading of the lighest area in a scene taken with a spot meter is used to locate that part of the scene at the most favorable point on the characteristic curve of the film. From another reading taken in the darkest area of the scene, the scene contrast may be calculated. This indicates to the cameraman whether or not additional lighting is required to obtain images on the film within the contrast limitations of the film or the television system in which the film will be reproduced.

There is as yet no standard speed rating method for use with spot exposure meters. This is not a serious disadvantage; a simple test exposure is all that is needed to establish the relationship between a given reading on the meter scale and a particular value of film density for a pre-determined setting of the camera lens aperture.

To make this test a three-step gray scale chart should be prepared with Munsell calibrated neutral papers, as shown in Fig. 41. The lightest area of the chart is Munsell neutral paper, Value N 8.0/, with reflectance of

* See *European Television Film Production Methods* by Adolf Hinze, *SMPTE Journal*, Jan. 1963, pp. 11–14.

approximately 60 per cent. The middle (gray) step is Munsell Value N 6.0/, with a reflectance of 30 per cent, half that of the lightest step. The black step is made from Munsell Value N 2.0/, with a reflectance of about 3 per cent. This chart represents a subject contrast range of 20:1.

This test chart should be placed in front of a camera and illuminated at a convenient level. (Only an approximately correct level of illumination for exposure of the film in the camera is needed at this stage.) Readings should be taken with a spot meter in the lightest and darkest steps of the test chart, and these numbers should be carefully noted for future reference. A series of exposures should then be made with the camera, over a range of lens

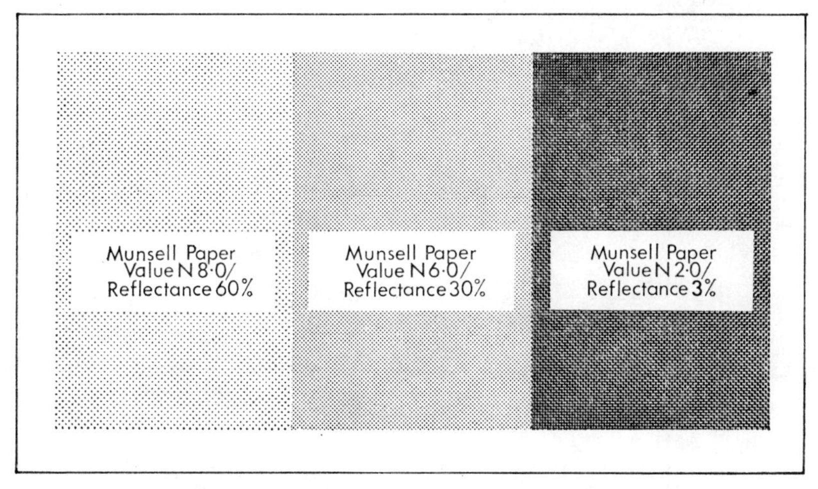

Fig. 41 Three-step gray scale test chart for setting up a television film exposure control system.

aperture settings (*f*-numbers). To identify these exposures, the lens setting for each exposure should be marked in a corner of the test chart. A short length of unexposed film is required at the end of the roll, for a sensitometric control exposure.

After processing, the densities of the sensitometric control strip should be measured and plotted on graph paper to obtain the characteristic curve. On this curve may be plotted the densities of the test chart, at each setting of the lens aperture. The lens setting which places the lightest and darkest areas of the test chart at the most favorable points on the characteristic curve provides the key information needed to set up an exposure control system with a spot meter.

In Fig. 42(a) is shown a typical characteristic curve for a color reversal film. Plotted on this curve are two points, S and H, taken from Fig. 38. These are the points from which the speed of reversal color films is calculated, in accordance with USA Standard PH2.21—1961. These are the most favorable points, too, at which the lightest and darkest areas of actual scenes should be reproduced. If, for a particular reading of the lightest area of the test chart, taken with a spot meter, it is found that the camera lens must be set at *f*8 to locate this area of the test chart at Point H, then it can be said that the spot meter is correctly calibrated for the particular type of color reversal film in use. With the camera lens set at *f*8, any scene area giving the same spot meter reading will also be reproduced at Point

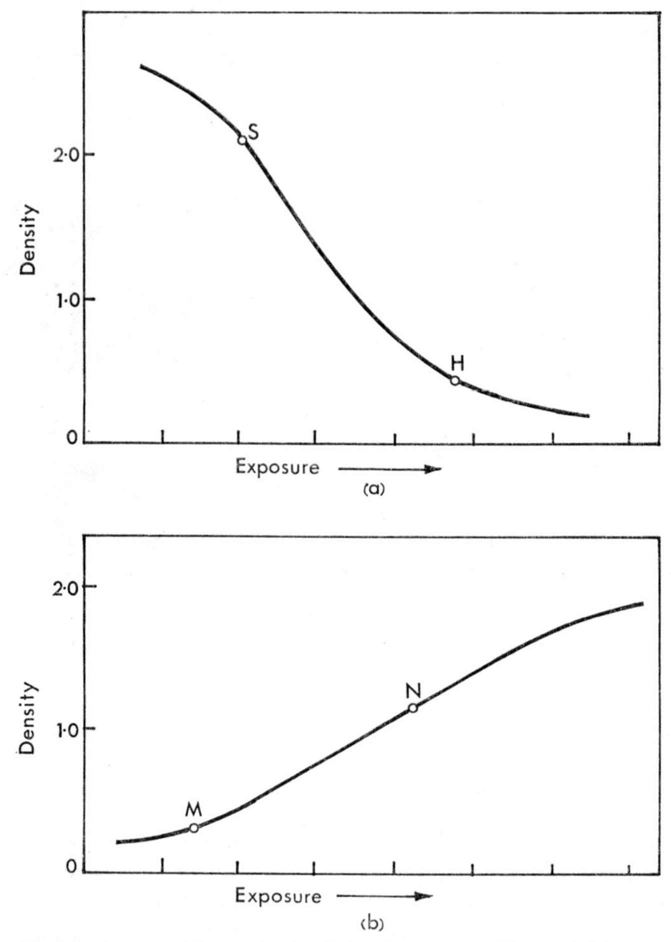

Fig. 42 Typical characteristic curves for (a) color reversal film and (b) color negative film. Plotted points show the most favorable locations on the curves for lightest and darkest scene areas.

H on the characteristic curve. For higher or lower spot meter readings, the lens aperture would have to be opened or closed, one camera lens stop representing an increase or decrease in exposure by a factor of 2.

A typical color negative characteristic curve is shown in Fig. 42(b). Here N and M represent the most favorable points for reproducing the lightest and darkest areas of the test chart—and the lightest and darkest areas of actual scenes as well. By making similar tests on all film stocks in use, a table may be made up and attached to the spot meter showing the meter readings required (at a given lens aperture setting) to expose each type of film properly.

In practice, many scenes will be encountered with contrast ranges far greater than the 20:1 range of the test chart. A typical situation is an outdoor scene with an area of bright sky and heavily shaded dark areas. Only rarely is it possible to make use of artificial lighting to illuminate the shadows and reduce the contrast. The most important consideration in a situation of this kind is to set the lightest area—the sky—at the correct point on the film's characteristic curve, so that, in subsequent reproduction in telecine, peak white signal levels will be properly maintained. This will result in the shadow areas of the scene being reproduced without detail.

When a scene such as this includes people's faces, it is likely that the luminance difference between sky and faces will be excessive. As a general rule, faces should be reproduced in the film at approximately the same level as the middle (gray) step of the test chart. If spot meter readings show that the luminance of the sky is more than about twice as great as the faces, it is quite likely that the faces will be much too dark relative to picture whites in telecine reproduction.

The cameraman has two alternatives. The simplest alternative is to alter the camera angle so that the bright sky area is excluded. This would permit some other less bright scene area to be selected as the peak white reference. As a result, the exposure of the film would be increased to locate this less bright area at the desired white point on the film's characteristic curve. At the same time, faces would be reproduced at a higher signal level in telecine, making these picture areas much lighter.

The other alternative is to make use of artificial illumination to reduce the brightness (luminance) difference between sky and faces. In this way faces—and other areas of primary picture interest as well—may be reproduced in the picture tonal scale at any desired level. A spot meter is particularly useful in making measurements of this kind, and in calculating the lighting levels required in different scene areas to obtain the desired pictorial effects.

6 Processing color films

Color films are made up of three light-sensitive layers, each of which responds to a different portion of the color spectrum—roughly red, green and blue. An object which reflects blue light, for example, produces an exposure only in the blue-sensitive layer; only the green-sensitive layer is affected by green light, and the red-sensitive layer by red light. In practice, of course, very few objects reflect light in such a way that only single layers of the film are affected—generally, the exposure in one layer predominates, with lesser amounts in the other two layers.

At the time of exposure, no color exists in a color film (except for the presence of colored couplers). The three layers contain light-sensitive silver halides which are affected by exposure in much the same way as a black-and-white film. Exposure makes the silver halide crystals susceptible to the chemical action of a developing solution. Thus, the first stage of color film processing is the same as for black-and-white processing: the silver halide crystals that have been exposed must be converted to black silver grains.

If at this stage, the three separate layers could be peeled off the base and examined, it would be found that all three layers contain black silver images with a generally similar appearance. However, careful comparison of the images would show that the amounts of black silver (density) in the three images are different, depending on the colors in the original scene that these images represent.

The next stage in color processing is to convert these black silver images into dye images, after which the black silver images must be removed. The procedure to be followed at this stage depends on whether the material being processed is a negative or positive, or a reversal film.

Processing Methods

For the past 40 years, it has been the practice to process motion picture film in continuous machines, with separate tanks for the processing solutions, and banks of rollers to carry the film through the solution tanks. A typical processing machine layout is shown in Fig. 43. From an elevator at the left hand end of the machine, the film is taken off a supply or feed reel, and drawn over the rollers in a series of helical loops.

Several different methods may be used to impart continuous motion to the film. The most common method is to pass the film over rollers fitted

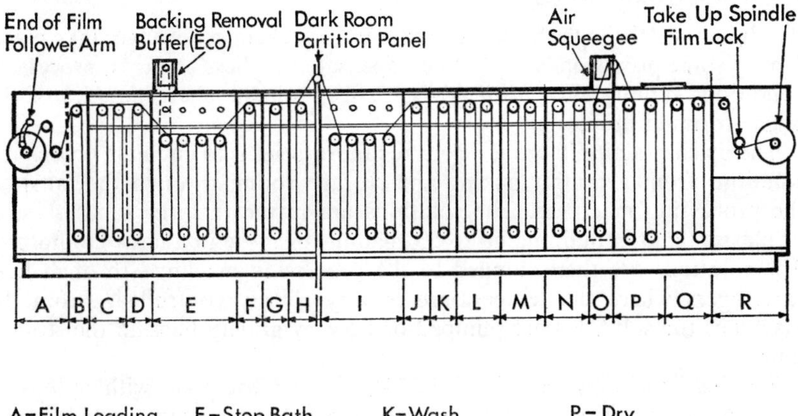

A–Film Loading
B–Feed Section
C–Prehardener
D–Neutralizer
E–First Developer
F–Stop Bath
G–Reversal Bath
H–Wash
I–Color Developer
J–Stop Bath
K–Wash
L–Bleach
M–Fixing Bath
N–Wash
O–Stabilizer
P–Dry
Q–Take-up Section
R–Take-up Reel

Fig. 43 Typical motion picture processing machine layout.

with sprocket teeth that engage the perforations in the film. These rollers are rigidly attached to shafts driven by a variable-speed motor.

Allowance must be made for the stretching and shrinking of the film as it passes through the machine, first in the processing solutions, and then in a drying cabinet where moisture is removed. One of the most severe problems in the design of a processing machine is to provide for constant film tension. If the film becomes slack, the strands may be entangled, causing a break, or physical damage to the delicate emulsion layers. On the other hand, excessive tension will damage the perforations and perhaps break the film.

To avoid these problems some processing machines have a tendency drive. Some of the rollers are fixed on positively driven shafts, and these rollers turn at a slightly greater rate than the film. In a machine of this type tension is maintained by friction between the film and the driven rollers. These machines require constant attention, for it tension is lost for any reason, slack film piles up in the machine, causing breaks and damage.

In the processing solutions the emulsion layers become quite soft, and are very easily damaged. The film path is arranged so that only the base side touches the rollers, and usually the rollers are cut away in the center so that only the edges of the film come in contact with the rollers. Machine designers have had some success with rollers fitted with rubber sleeves over which the base of the film rides. Rollers of this type support the film over its entire width, minimizing physical distortion of the base, especially when excessive tension develops in the machine.

Solution Handling

As film passes through the processing solutions chemical changes take place. Unless some provision is made to compensate for these changes, processing conditions rapidly alter, with adverse effects on the characteristics of the image-forming process.

The usual practice in motion picture laboratories is to circulate processing solutions from a central mixing and storage room, through the tanks in the processing machines. The solutions are mixed in large stainless steel or plastic tanks containing up to 500 gallons or more in a large laboratory. The solutions are then pumped to the processing machines through heat exchangers where the temperature is very closely controlled. From the machines the solutions are pumped or flow by gravity back to the storage tanks.

Smaller processing machines are likely to be equipped with pumps to circulate the solutions within the machine tanks. In some machines the solutions are forced through jets located close to the film strands, to promote vigorous agitation of the solutions at the film surfaces, and uniform chemical reactions.

In most laboratories it is customary to add replenisher solutions to the main storage tanks, to compensate for the changes in chemical composition that take place due to the reactions between the solutions and the film. The replenisher solutions usually have a composition different from that of the original starting solutions, and the rate of replenishment must be adjusted to balance exactly the losses occurring in these solutions. In some cases, reactions between solutions and film release chemical substances into the processing solutions, and the concentration of these substances tends to increase. To counteract changes of this kind, the main solutions must be diluted by the replenisher solution, which, of course, should not contain any of the substance that is being released.

Control of Processing Solutions

In the earlier black-and-white era, chemical control was largely a matter of trial and error, except in the largest motion picture laboratories. With the advent of color, every laboratory, even the smallest, was obliged to adopt chemical control procedures, to ensure uniform color image formation. Moreover, it is essential in the processing of color film to adhere closely to the recommendations of the manufacturer of the film. Unless this is done, the quality of the color images is likely to be unsatisfactory, and this can easily lead to serious losses.

Compared to the processing of black-and-white film, color processing is quite complex. The control of processing includes four distinct and essential phases:

1. A knowledge of the desired characteristics of the process and the end product is essential.
2. The status of the process must be continuously evaluated to determine whether or not these characteristics are being maintained.
3. When the evaluation indicates that these characteristics are not being maintained, a diagnosis must be made to determine the cause.
4. The proper corrective action must be taken.

Process specifications are normally set by the film manufacturer, and supplied to the laboratory in the form of instructions for chemical mixing, handling and control. When these procedures are followed, the characteristics of the color images obtained with the type of film in use should be satisfactory.

To ensure that process specifications are being met, the user must set up an evaluation procedure. Sensitometric control strips processed at regular intervals will show deviations from normal. To provide a continuous evaluation of the process, measurements must be made on the processed strips and the results plotted on a control chart.

Variations in the densities of the control strips will show that changes are taking place in the processing conditions, but this method of evaluation cannot provide any indication of the reasons for the variations. To obtain this information, the processing solutions must be analyzed to determine the nature and extent of the changes that have taken place in the composition of the solutions.

Sensitometric Control Procedures

It is a quite simple undertaking to make a sensitometric analysis of a color processing condition. To make such an analysis two relatively inexpensive instruments are needed—a sensitometer and a densitometer. The sensitometer is used to make exposures on short strips of film, known as control strips, and after processing, measurements are made on these strips with a densitometer.

A sensitometer consists basically of a light source, an exposure modulator, a shutter, and a device to hold the film in the exposure plane while the shutter is actuated. The simplest form of exposure modulator is a gray scale step tablet or step wedge, with 21 steps, in which density increases by a factor of 0.15. The light source must be very carefully controlled to ensure constant light level at the exposure plane, and to avoid the effects of any changes in color temperature.

The basic elements in a densitometer are a light source, a photocell and a

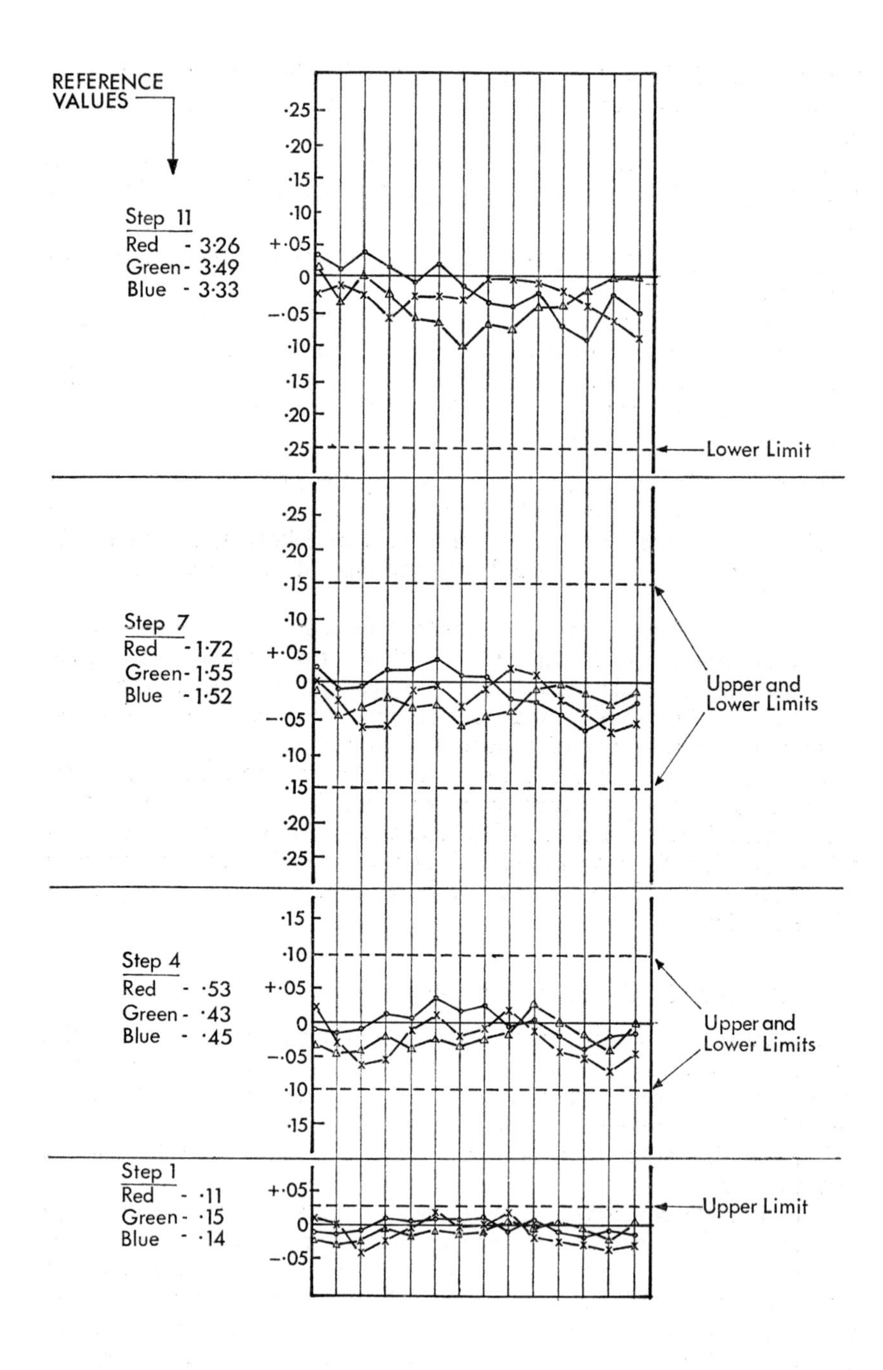

Fig. 44 Color processing control chart.

microammeter calibrated to give readings in density values—that is, in terms of the common logarithm of the reciprocal of the transmittance. In a color densitometer, three color filters are used, to enable measurements to be made of the density of the film to red, green and blue light. In this way, the characteristics of the dye images may be analyzed, and displayed in the

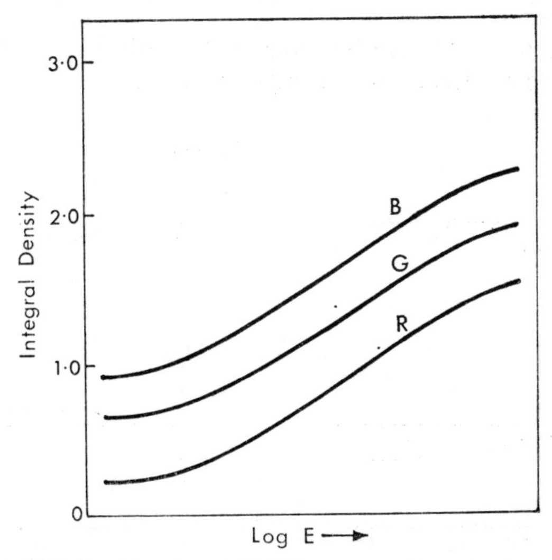

Fig. 45 Characteristic curves for a color negative film with incorporated colored couplers.

form of three curves, as shown in Fig. 45. These curves represent the red, green and blue integral densities* of a color negative film.

In addition to or in place of gray scale wedges, color control strips may have patches of single colors, the exposures being made with narrow-band filters affecting single layers of the film. In some cases, arrangements may be made with the film manufacturer to provide supplies of exposed control strips. This avoids the need for maintaining precise sensitometer exposure conditions in individual laboratories.

Chemical Control Procedures

When film is processed halide ions are released into the developer, some of the developer constituents are used up or chemically changed, and organic development by-products are formed. At the same time, carry-in of previous solutions or washes tends to dilute the developer, while carry-out of the solutions on the film tends to reduce the volume.

* The term 'integral density' is defined as the effects of image absorptions on a receiver in an optical system. There are several types of integral density, depending on the kind of response that is controlled by the image or the kind of receiver in which the response is generated. For further information on this subject see *Principles of Color Sensitometry* published by the Society of Motion Picture and Television Engineers, New York, N.Y.

To maintain a satisfactory level of chemical composition and volume, proper replenishment is required. The replenisher must be formulated and added at a rate that will:

1. Replace chemicals used up or chemically changed.
2. Maintain the concentration of halide ion and development by-products at suitable levels.
3. Compensate for dilution by carry-in of solutions.
4. Replace tank solution lost through carry-out.

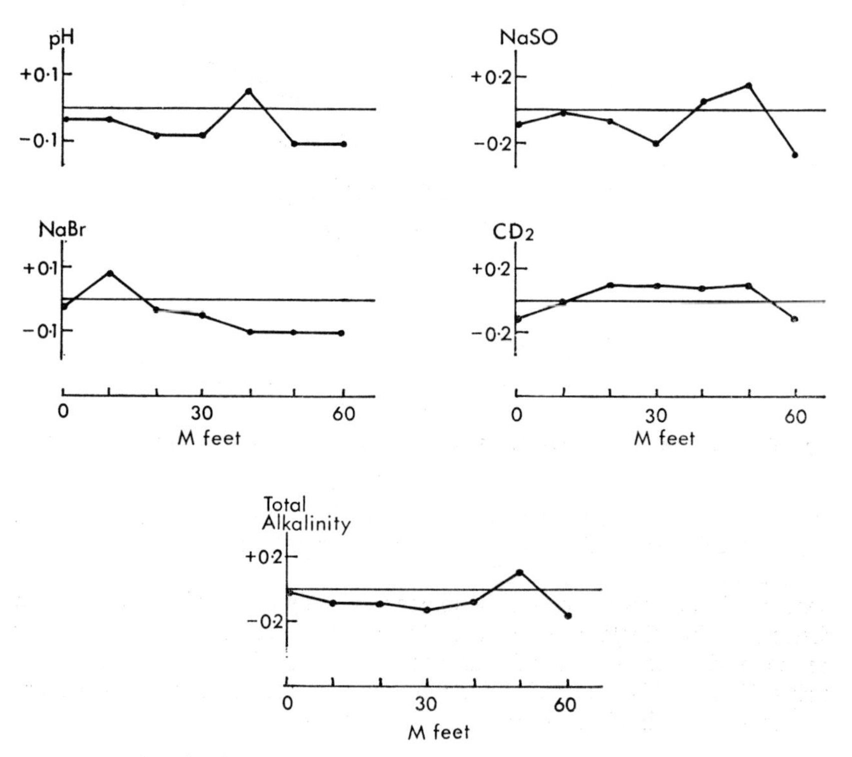

Fig. 46 Chemical analysis control chart for a color developer.

The volume of replenisher added to the developer system determines the volume of solution overflowed by displacement. It is wasteful to use a greater volume of replenisher than necessary. The minimum volume of replenisher that can be used is determined by the required concentration of the least soluble chemicals and the volume of tank solution that must be displaced to restrict the build-up of halide ions and organic by-products to suitable equilibrium levels.

While careful attention to prescribed procedures for replenisher composition and rate of flow usually ensure uniform processing conditions over short periods, chemical analyses of the tank solutions provide additional information vitally important for ascertaining the cause of trouble when sensitometric data show that color image forming conditions have changed. To make these analyses the services of a qualified chemist are required.

Processing Color Negatives

In professional 35 mm operations it is customary to use color negative film in the camera. Afterwards, any desired number of color prints may be made from these negatives.

The images in a processed color negative are complementary in color to the objects in the original scene—that is, a red dress is reproduced as a cyan (blue-green) area. When a copy (print) is made from the negative on

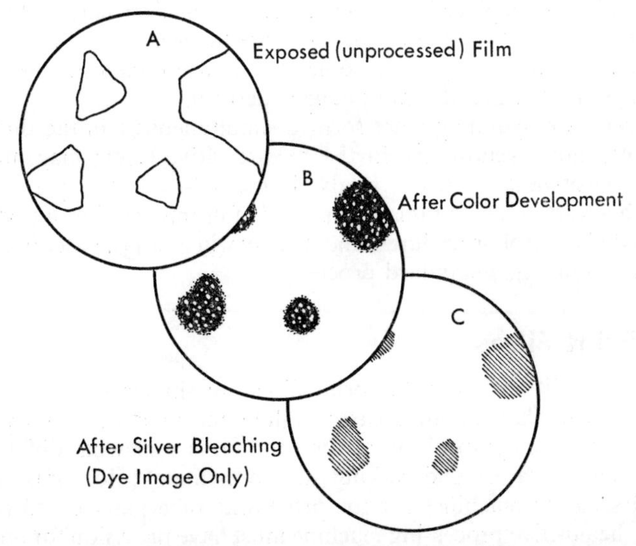

Fig. 47 Color image formation (a) silver halide crystal before color development (b) silver grain after color development (c) after silver removal by bleaching.

color print film, the cyan area in the negative is reproduced as a red area in the print, similar in appearance to the red dress in the original scene.

Most color negative films have colored couplers incorporated in the layers to help compensate for the imperfect light absorbing and transmitting characteristics of available dyes. The presence of residual colored couplers in processed color negatives gives these films a typical orange cast.

Color films usually have a black anti-halation backing to reduce light reflection within the film at the time of exposure. This backing must be removed during processing. As a rule it is softened in the alkaline first developer, after which it can be flushed off with a water spray.

Next, the film passes through the color developer solution, where silver and dye images are formed simultaneously, as shown in Fig. 47.

Color development depends on the presence in the developer or the emulsion of substances known as couplers.* The developing agent in the developer solution reduces exposed silver halide crystals to black silver grains and in doing so becomes oxidized. When a coupler compound is present, the oxidized developing agent reacts with the coupler to form an insoluble color dye.

This reaction takes place only with some developing agents such as paraphenylenediamine and some of its derivatives. The dye is formed from the coupler and the oxidized developer, and the color of the dye formed depends on the nature of the coupler and the developer used. The amount of dye formed depends on the amount of oxidized developer available, and this in turn depends on the amount of silver developed; thus the amount of dye is related to the degree of exposure which each point in the film receives.

The reaction takes place in the neighborhood of silver halide crystals that have been exposed to light, and dye spots are formed around the developed silver grains, reproducing in a somewhat blurred manner the granular silver image from which the dye image is derived.

Three separate dye images are formed simultaneously in the three layers of the color film—yellow dye in the blue-sensitive layer; magenta dye in the green-sensitive layer, and cyan dye in the red-sensitive layer.

Manufacturers of color films will provide full information on processing solutions and control procedures, and usually will assist the user in establishing and maintaining a standard process.

Color Print Films

Processing conditions for color print films are similar to those for color negatives except that, in some cases, a different color developing agent is employed, and the composition of the solution is somewhat different. This means that a laboratory undertaking the processing of color films, may have to install separate machines for the processing of negatives and prints. In addition the positive processing machine must have provision for developing the optical sound track (see p. 102).

One of the most important advantages of a color negative-positive process is that the negative material yields relatively low contrast images. This provides greater latitude for exposure errors in the taking camera. Afterwards, when prints are being made on relatively high contrast color print film the contrast of the images is increased sufficiently to produce pictures with satisfactory appearance on projection screens.

Because of the importance of precise processing control for both color negative and print materials, it is not practical to attempt to vary picture contrast by adjustment of processing conditions. Preferably, contrast should be controlled by adjustment of lighting conditions in the original scene, at the time of the exposure of the negative.

* For futher information on this highly complex subject see *Chemistry and Color Photography*, a tutorial paper by P. W. Vittum in the *SMPTE Journal*, December 1962.

Variations in the color balance of the pictures due to film and processing variables may be corrected in printing the color negatives. This subject is dealt with in greater detail in Chapter 8.

Color Reversal Films

Reversal films achieved great popularity in the black-and-white era, especially in amateur and semi-professional applications of the 16 mm format. For many purposes, the original film exposed in the camera, processed by chemical reversal to yield positive picture images, was the only copy required.

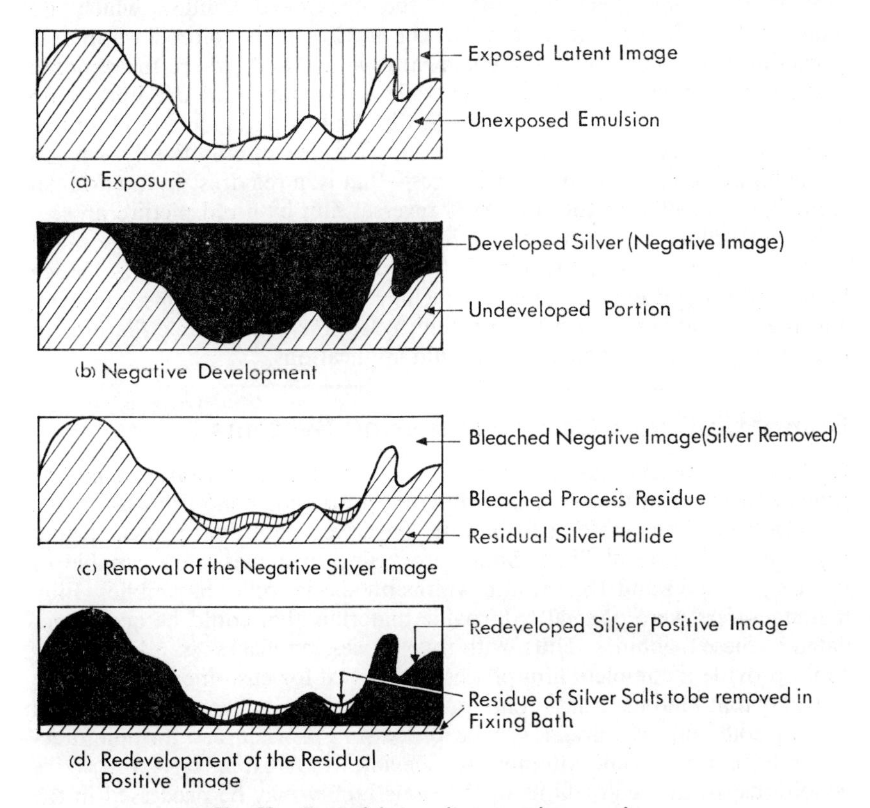

Fig. 48 Essential steps in reversal processing.

With this procedure, users avoided the time and expense of making prints. The quality of the pictures obtainable with properly exposed and processed reversal films is excellent—at least as good as, and in most cases much better than, prints made from 16 mm negatives.

The first three-layer color film made available to the public was a reversal material, and these films were used extensively up to about 1950 when the first color negative and positive materials appeared on the market. Reversal films are still used for most 16 mm applications, while in 35 mm professional motion picture work the negative-positive system is employed.

When it is necessary to make copies of reversal originals two courses may be followed. The simplest and most direct is to use a reversal print film for the copy. Alternatively, an intermediate color negative may be made from the reversal original, and then the desired number of prints may be turned out using color print film.

In the processing of reversal color film, the exposed halide crystals are first developed to black silver. Then, after a wash and acid rinse, the film is given an overall exposure to light or is passed through a chemical solution to provide the same effect—fogging of the unexposed halides, which are unaffected by the first development. The next stage is color development (sometimes combined with the chemical fogging step), where the exposed crystals are converted to black silver while at the same time dye images are formed. Finally, the silver grains formed in the first and color development stages are removed in a bleaching solution. This leaves only the dye images in the film, and these are positive images—that is, a red dress in the original scene is represented in the processed reversal film by a red picture area.

Reversal films, having relatively high contrast characteristics, are much more sensitive to exposure errors than color negative materials. Much higher speeds and finer grain can be achieved with reversal films, however, and these advantages, to a large extent, offset the unfavorable features of the reversal system, especially in 16 mm applications.

Compatibility of Color Processing Systems

In the earlier black-and-white period, motion picture laboratories had to provide three, or at the most four different processing conditions to handle all available film materials—picture negatives, positive prints, sound negatives and reversal films. Some processing machines were capable of handling both 35 and 16 mm film widths on special roller assemblies. Film manufacturers were obliged to provide materials that could be accommodated in these machines. Thus, with four processing machines, a laboratory could provide a complete film processing service for customers.

Today that kind of situation no longer exists. With a few exceptions, each type of color film requires a special processing procedure. Film manufacturers have made some attempts to alleviate these exceedingly restrictive conditions, so that color films of the same family may be processed in the same machine without changing the tank solutions. Films of the same type from different manufacturers—for example, color negatives—require different processing solutions and cannot be processed in the same machine. This severely limits laboratory interchangeability.

In 1961 the Eastman Kodak Co. demonstrated a processing machine of unique design, in which viscous solutions were applied to the surface of the film.* This technique, the first major advance in film processing technology

* See the November 1961 issue of *SMPTE Journal*, pp. 875–881, for two papers on *Rapid Processing of Motion Picture Film by the Application of Viscous Coatings*.

in 40 years, appeared to offer a number of important advantages over conventional deep-tank machines. Its possibilities in the processing of color films were investigated but no commercial applications have yet been developed.

Viscous layer processing would permit the design of small machines, capable of operation in normal room light. Change-over between different types of processing solutions to accommodate a variety of color films would be quite simple, requiring only the interchange of solution supplies from one set of containers to another.

Simplification of Color Processing

In response to the demands of television for simpler methods of color processing, considerable advances have been made within the past few years. With the introduction in 1965 of high-speed Ektachrome reversal color films by Eastman Kodak Co. (Types 7241, daylight, and 7242, tungsten balance), a modified processing technique, known as the ME4 process, was announced, operating at an elevated temperature of 100°F, to reduce processing time by about two-thirds. This permits the use of smaller, simpler and less expensive processing machines. In addition, the film manufacturer supplies pre-packaged processing solutions, and detailed instructions for their use, to enable users to maintain uniform processing conditions.

Many television stations have already installed small, self-contained processing machines for reversal color films. These processors can be operated by personnel with limited motion picture training and experience.

A paper on continuous film processors in the March 1963 issue of *SMPTE Journal* (page 184) provides considerable background information on the design and construction of these machines.

7 Color printing and duplicating

In its simplest form color printing consists of exposing a color positive material in contact with a color negative, using a white light source for the exposure. The operation is carried out in a continuous contact printer in which the color negative controls the exposure of the print film. Prints may be made from a reversal original in a similar manner, except that the original has positive images, and a reversal material is used for the prints, to again obtain positive images.

Two methods of color printing are commonly employed. The first, known as subtractive printing, uses a white light source such as a tungsten lamp. The exposure of the three layers of the print film is modified by inserting color correction filters in the light path. These filters absorb or subtract small amounts of one or more of the colors from the white light.

The second method is known as additive printing. Here, instead of a single white light source, three colored sources—red, green and blue—are mixed to produce light of any desired color. Additive printing requires a specially constructed machine, whereas a conventional black-and-white printer, modified for the insertion of correction filters, may be used for subtractive printing.

Most color film productions require a duplicating stage of some kind. For example, when 35 mm color negatives are used in a production, it is often necessary to make 16 mm reduction prints for distribution. First a 35 mm intermediate positive is made from the edited color negative. Next, a reduction negative is made on 16 mm color intermediate film, and finally, the required number of prints are made with 16 mm color print film.

To introduce effects such as fades and mixes in a program, portions of the original color negatives on either side of each effect must be copied in an optical effects printer on color intermediate film. Effects in 16 mm productions are achieved usually by a simpler A & B roll printing technique, but titles superimposed over scenery may call for one or more duplicating stages.

Color duplicating is an extremely critical type of work, mainly because errors in exposure level or in color balance at any stage will result in severe distortion of the color rendering in the prints.

Motion Picture Printers

The standard rate of film movement in cameras and projectors is 90 ft per minute for 35 mm and 36 ft per minute for 16 mm. When large numbers of prints must be made for distribution it is highly desirable to speed up the rate of printing to several times these figures. This can be done by running

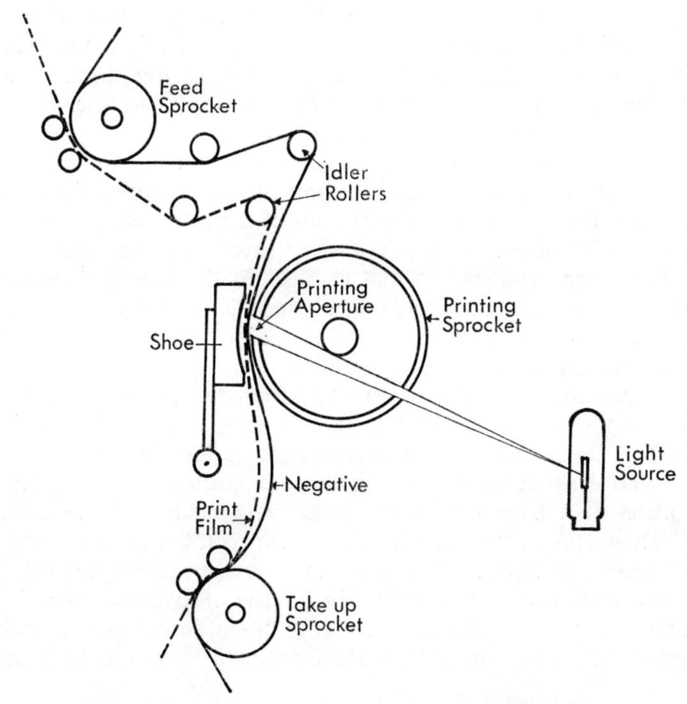

Fig. 49 Basic operations of a continuous contact printer.

the negative and positive stock in face-to-face contact over an illuminated slit, as shown in Fig. 49.

A large sprocket engages the perforations in the two films, and draws the films at a constant rate across the slit. A curved gate helps to conform the films into a path that follows the sprocket for a portion of its circumference. While the films are passing over the slit, light from the printer lamp is modulated by the images in the negative film to form latent images in the print film. The width of the slit or the intensity of the printer light may be adjusted to vary the exposure of the print film. Excellent results can be achieved with this method of printing up to speeds of 200 or 300 ft per minute.

Subtractive Color Printing

Only minor modifications are needed to adapt a contact printer for color film operation. Some means must be provided to alter the color of the printing light to compensate for variations in the color balance of the camera originals from which the prints are being made. At the same time the intensity of the printer light has to be adjusted so that the prints will have uniform brightness scene-to-scene when projected on a screen.

The usual procedure in subtractive printing is to make up packs of color compensating (CC) filters for each scene change. These filters absorb partially in one region of the spectrum and transmit freely in the remainder, permitting wide variations in the color of the printer light. Additional neutral density filters may be added to the filter packs to alter the level or intensity of the light level at the printing slit.

The term subtractive color printing is derived from the use of these color compensating filters to subtract portions of the white light from the printer lamp. The overlapping absorptions of the color compensating filters lead to some difficulties. Ideally, these filters should have spectral absorption curves with steep gradients so that changes in the energy distribution of the printing light could be effected over a specific band-width related to the spectral transmission characteristics of the color negatives and the spectral sensitivity characteristics of the print film.

It is not possible to produce filters with these ideal characteristics. Combinations of filters, especially when several have to be used to make up a pack, lead to losses in color contrast and saturation. Besides, it cannot be assumed that the removal of a particular filter from a pack is equivalent to the addition of a filter with a complementary color and the same peak density. Then, too, neutral density filters used with the compensating filters to keep the overall light level constant may not be entirely neutral. Such a system obviously becomes inefficient in the use of available light. There is the problem, too, that color compensating filters do not remain stable for long periods of time in the path of the high intensity printer light source.

Color Compensating Filters

Color compensating filters are available in yellow, magenta, cyan, red, green and blue, in seven different densities as follows—

CC025 CC05 CC10 CC20 CC30 CC40 CC50

The number of these filters, divided by 100, indicates the average density of the filter in the portions of the spectrum where the filter absorbs light to the greatest extent.

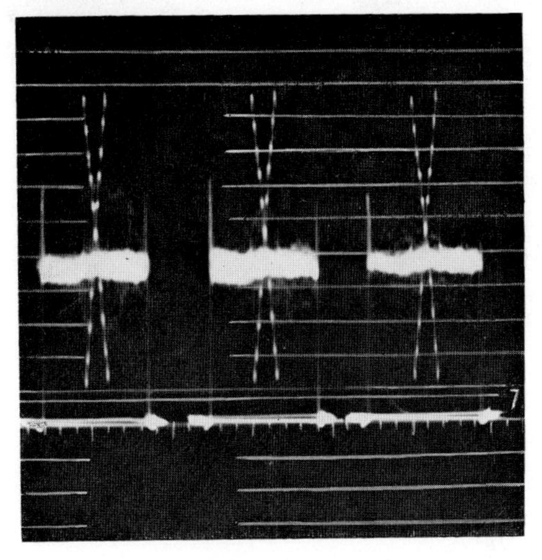

Plate 1—Waveform monitor display showing the output of a three-tube color camera reproducing a stepped gray scale test slide.

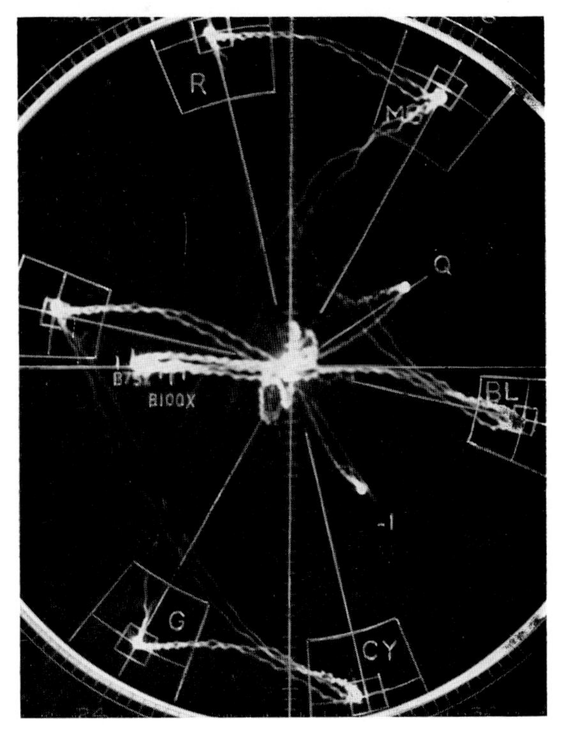

Plate 2—Vectorscope pattern obtained from a color bar test signal, with a correctly adjusted encoder.

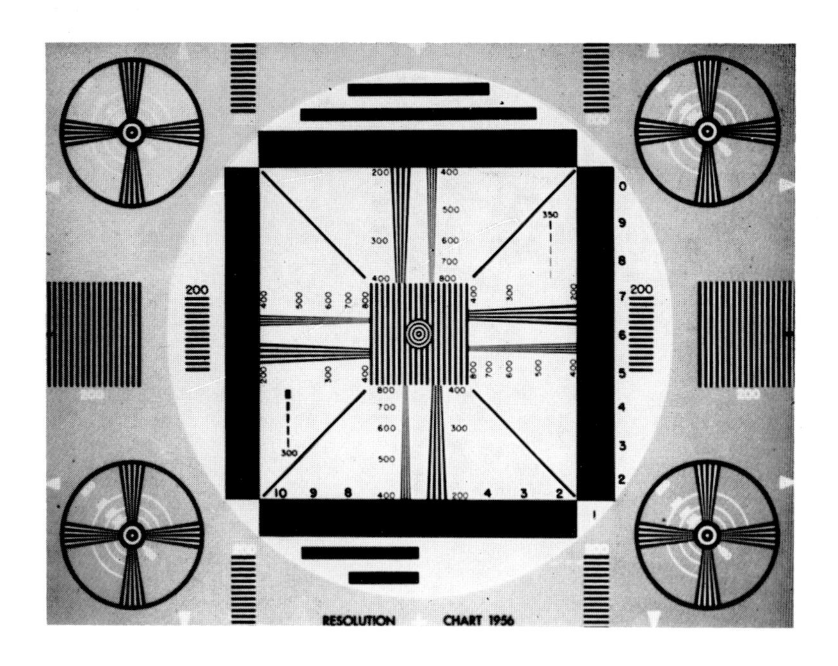

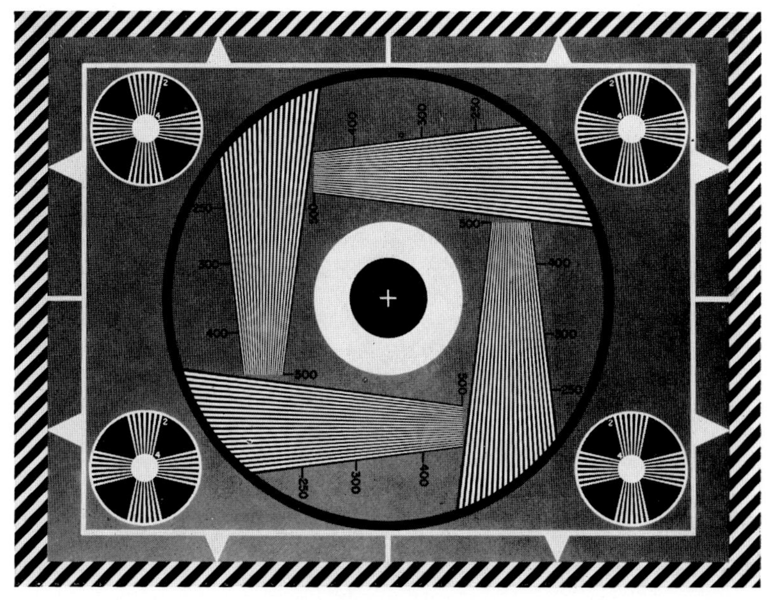

Plate 3—Top—EIA resolution slide. Bottom—resolution and alignment section of SMPTE television test film.

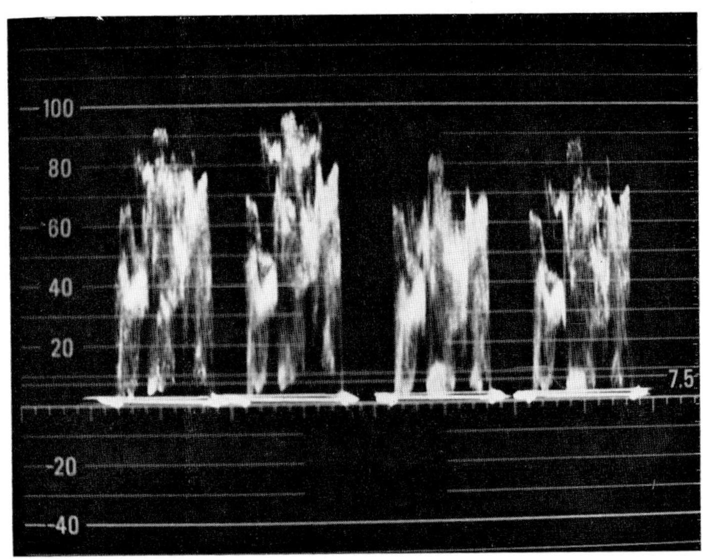

Plate 4—Waveform display from a color film picture in a four-tube color camera.

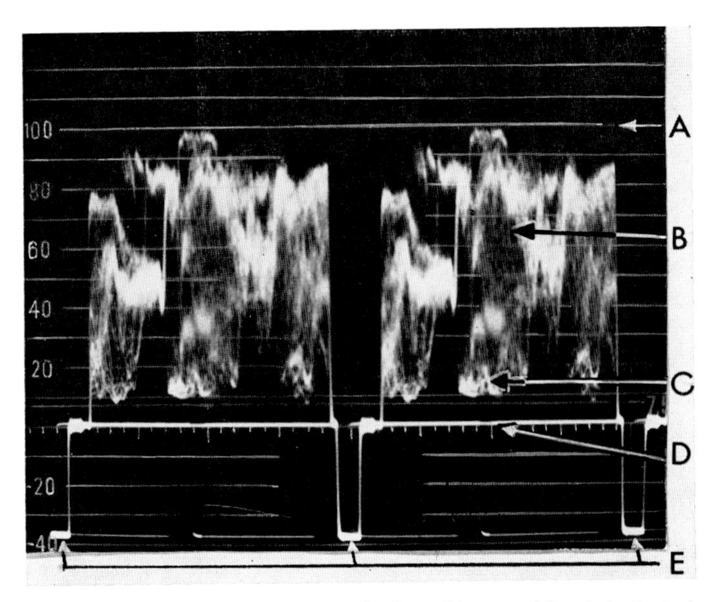

Plate 5—Typical video waveform display with several levels indicated: (a) white level (b) face tones (c) black level (d) blanking level and (e) horizontal sync pulses.

For each batch of color film, the printer must be balanced to compensate for slight differences in color characteristics. Sensitometric exposures and practical picture tests will indicate the most favorable combination of CC filters to correct for these differences.

If these tests show, for example, that the results obtained with one batch of print stock are more yellow than the previous batch, a yellow compensating filter would be added to the basic filter pack. The density of the filter added depends on the amount of correction required. Adding a yellow filter reduces the amount of blue light reaching the print film through the filter pack.

This results in less exposure of the blue-sensitive layer, and when the film is processed, less yellow dye is formed. The amounts of dye formed in the other layers can be controlled in a similar manner, by adding or subtracting appropriate CC filters in the basic filter pack.

To alter the color balance of individual scenes in a film, the color of the exposing light must be changed in a sufficiently short time to avoid perceptible changes, scene-to-scene, in the pictures in the print. Ideally, the change should be made within the frame line between scenes, but in practice a one-frame tolerance is about the best that can be achieved.

Many different methods have been employed to alter the color of the illumination in subtractive printers. One method consists of a filter changer working on the principle of a slide projector. Individual filter packs are made up and mounted in holders and stacked in sequence in a feed magazine. A light-changing mechanism in the printer activates the filter changer, bringing a filter pack into the light beam. For every change in exposure level or color that is needed during printing, a filter change takes place.

Additive Printing

In the additive printing method, red, green and blue light of suitable spectral composition is obtained from three separately filtered sources, or from a single source divided into three separate filtered beams. These separate colored lights are then combined at the printing aperture in any desired proportions. Modulation of the level or intensity of the three beams is effected by vanes, diaphragms or neutral density filters actuated by notches in the edge of the negative film, or by patches of foil attached to the film, or by paper or magnetic tapes advanced synchronously with the negative film in the printer. In this manner both color balance and exposure level may be adjusted as required.

The additive printing method is superior to subtractive printing because spectral bands may be selected to provide the most favorable reproducing conditions for any color printing system. By making the spectral bands very narrow, the effects of undesirable absorptions and transmissions of practical film dyes may be minimized. Significant improvements in color saturation and fidelity can be achieved in this way.

Additive printing requires more complex and costly equipment, but the time-consuming preparation of subtractive filter packs is eliminated. Light control in an additive printer is effected electromechanically, once the three spectral bands have been selected.

Many different methods have been used to select and control the three spectral bands. Fig. 50 shows one method. The light from a high-intensity

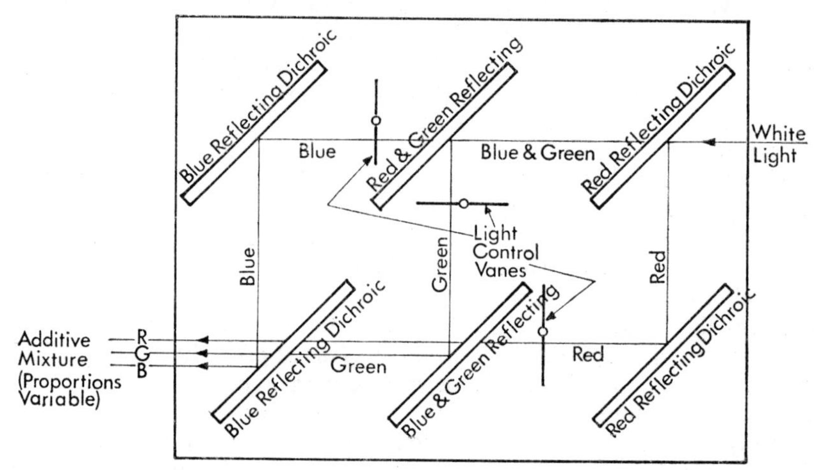

Fig. 50 Light-control system in an additive printer, and the paths of the colored light beams.

tungsten lamp is split into three beams by dichroic mirrors. In each color beam there is a movable shutter vane which can be variously located by the control mechanism in the printer. Another set of dichroic mirrors is then used to recombine the color beams at the printing aperture.

Colorimetry of Additive Printing

Ideally, the cyan dye in a color negative should transmit only blue and green light; the yellow dye should transmit only red and green, and the magenta dye, blue and red. In other words, the densities of the dyes should be as low as possible in these portions of the spectrum. At the same time, the density of the cyan dye to red light, the yellow dye to blue light, and the magenta dye to green should be as high as possible. This would ensure that no light of these colors passes through the dye layers.

The spectral density curves for a typical color negative film in Fig. 31 show that these ideal conditions cannot be achieved in practice. All of the dyes have some density in portions of the spectrum where light should be transmitted freely, while some light is transmitted in regions of the spectrum where density should be at a maximum.

When prints are made with a white light source, as in a subtractive printer, the undesirable absorptions and transmissions of the dyes in the negative give rise to images in the positive film that do not have the correct relationships with the images in the negative. These degradations and distortions of colors can be reduced by printing with a mixture of three narrow spectral bands, selected to coincide with the portions of the spectral density curves where density is greatest for light of unwanted colors, and lowest for the desired colors.

In selecting three spectral bands for additive printing it is necessary

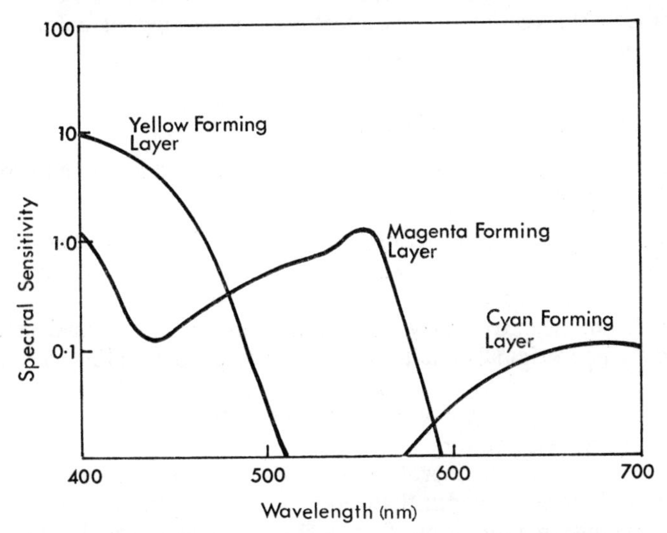

Fig. 51 Spectral sensitivity curves for a color print film.

to take into account not only the characteristics of the negative dyes, but also the spectral sensitivity characteristics of the positive print film. Ideally, the spectral bands selected for printing should coincide with the peaks in the spectral sensitivity curves of the print film (see Fig. 51).

Printing Reversal Color Films

Printing of reversal films is in some ways much simpler than printing negatives. In the first place, reversal originals have positive color picture images that can be viewed directly with the eye, and prints made from these originals can be compared to evaluate the nature and extent of any degradation or distortion that may have occurred in the printing process. A color negative has little visual significance when viewed with the eye, because colored objects are reproduced in their complementary colors. Besides, most color negative materials have incorporated colored couplers giving the negative images an overall orange cast.

The use of reversal color films is limited mainly to the 16 mm format. The production of 16 mm color films for educational, advertising, training and documentary purposes was at first a by-product of the introduction in 1935 of reversal color films for home movie making. In the late 1930's, professional use of these amateur-type films had reached such an extent that reversal print films were made available in bulk form to make better and lower cost prints from the originals. Later developments led to the introduction of a considerable variety of new 16 mm films, mostly of the reversal type. Some of the original camera films are not intended for direct projection, but rather for the making of prints.

Reversal color originals may be printed by the subtractive or additive methods. All the considerations affecting the printing of color negatives apply equally to reversal originals.

Techniques in Color Duplicating

The reasons for duplicating in motion picture production may be summarized briefly as follows—

1. Finished productions require a number of special effects such as fades, dissolves, wipes, etc., which are achieved most conveniently by using duplicating stages.
2. Duplicate negatives may be required as insurance against loss of valuable camera originals due to physical damage, dye fading, or improper storage.
3. Duplicate negatives are needed frequently for shipment to other locations for release printing.
4. Reduced or enlarged negatives are required when the camera originals are larger or smaller in width than the final release prints.
5. In making 16 mm prints from color reversal originals, color internegatives offer economic advantages where quantities of prints are required in comparison with making prints on another color reversal film.

The most important requirement of a duplicate negative is that it should have suitable characteristics for making prints as close as possible in appearance to prints made from the original negative. This means that the tonal gradation, contrast, hue and saturation of colors, definition and graininess in the print from the duplicate negative should be similar to a print from the original negative.

In duplicating work, sensitometry is an essential tool. Sensitometric control strips are used to monitor the color processes involved in the duplicating system. A basic rule in the control of duplicating processes is that only the central straight line portions of the characteristic curves should be used. This ensures optimum tone reproduction—a one-to-one transfer of tonal values.

In recent years film manufacturers have made available a special type of material for duplicating known a color intermediate film. This material may be used for making color intermediate positives from original color negatives, and also for making color intermediate negatives from color intermediate positives. Fig. 52 shows characteristic curves for a typical color intermediate film. These curves, indicating density to red, green and blue light, are separated by the density contributed by the colored couplers incorporated in the film layers.

The control of two stages of color duplication—intermediate positive and intermediate negative—is quite difficult. If, for example, an error is made in printing the intermediate positive, it is not likely that the fault will be discovered until the intermediate negative has been made, and a print from this negative is screened. The time and expense involved in remaking the intermediate positive, another intermediate negative and print can seldom be justified unless the fault is very serious.

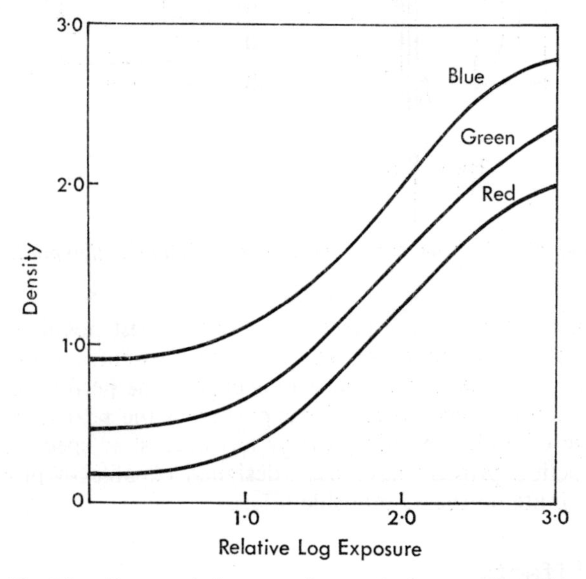

Fig. 52 Characteristic curves for a color intermediate film.

Recently, color reversal intermediate materials have become available, permitting a duplicate negative to be made from an original negative in a single step. This type of film can be obtained in both 35 and 16 mm formats.

Optical Reduction Printing

The demands for optical reduction printing have increased greatly in recent years, due mainly to the very heavy usage of 16 mm films in television programming. Professional film producing organizations prefer to use 35 mm for the camera originals, even when only 16 mm prints are needed for distribution.

Optical reduction requires a lens between the 35 mm original and the 16 mm material on which the smaller picture images are formed. The reduction printer can be thought of as being made up of a projector at one end and a camera at the other, with a lens so positioned that an image of the 35 mm film frame in the projector gate is sharply focused in the correct reduction ratio at the plane of the film in the camera gate.

Machines for optical printing usually advance the films one frame at a

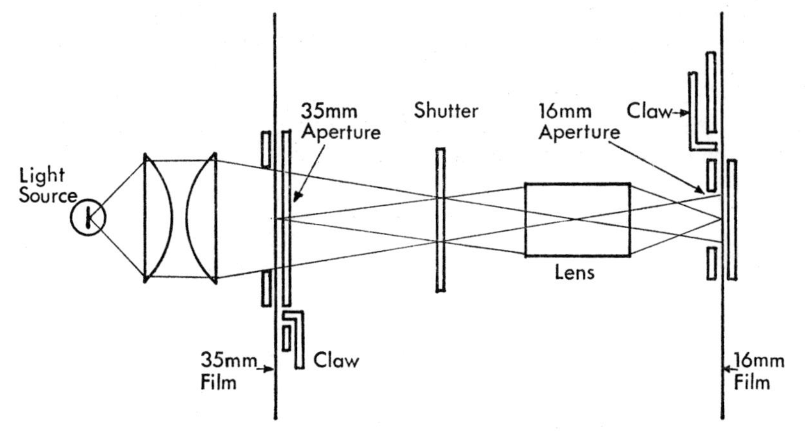

Fig. 53 Essential components of an optical reduction printer.

time. Often, register pins are provided to ensure best possible registration and steadiness in the reduction, and to avoid movement between the two films at the time of exposure. These pins engage the perforations after the films have been advanced ready for exposure of the next frame. A major disadvantage of optical printing always has been slow speed, but recently optical reduction printers have been designed capable of printing up to 200 ft per minute on the 35 mm side.

Optical Effects

Optical effects printers are much more versatile, incorporating features such as the following:

1. Automatic follow-focus controlled through a lead screw covering a range from 1:3 blow-up to 4:1 reduction.
2. Automatic dissolves from 8 to 120 frames in length.
3. Interchangeable film transport (projection head) for 35 and 16 mm films.

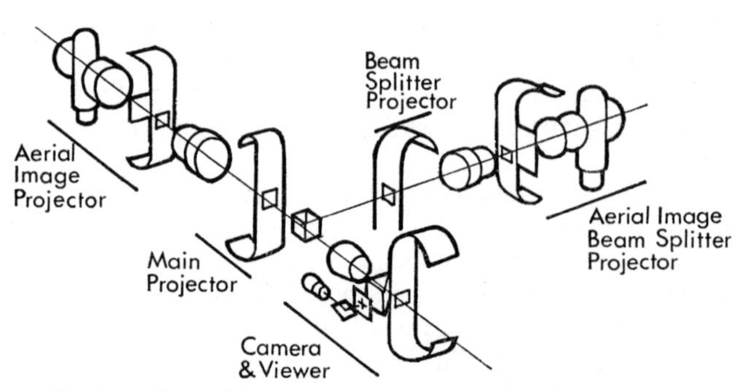

Fig. 54 Essential components of an optical effects printer.

4. Tilting mechanism to align scene objects in the film frames.
5. Precision movements on lens mount and projector head accurate to 0.001 inch.
6. Variable speed drive from stop-motion to 240 rpm.
7. A clutch mechanism to produce skip-frame effects.
8. Provision for use of anamorphic and other special lenses.

With the aid of an optical printer of this type, a great variety of the most ingenious effects can be produced, limited only by the imagination of the producer and the skill of the operator.

8 Color balance control

One of the most important advantages in the use of color film is the ability to modify the appearance of the color pictures when copies (prints) are being made. This can be done very simply by altering slightly the color of the printing light, so that the layers of the print film are exposed differently. Let us say that the original is a reversal film, with positive color images, and the appearance of these images is too yellow. The pictures in a reversal duplicate of this original film can be made more pleasing (less yellow) by placing a blue filter in the printer light beam.

This reduces the amount of yellow light reaching the duplicating material —that is, the exposure of the red- and green-sensitive layers is reduced, relative to the blue-sensitive layer. As are sult, greater amounts of magenta and cyan dyes are formed in these layers during processing, providing a better balance relative to the yellow dye image formed in the blue-sensitive layer.

A similar procedure is followed in altering the color balance of positive prints made from color negatives, but here the control of color balance is considerably more difficult.

In the first place, the images in the color negative are reversed—that is, the image of a red dress is cyan in color. Then, too, it is customary to incorporate colored couplers in color negative materials to compensate for dye deficiencies, and these impart a distinct overall orange cast to the film. As a result, it is impossible to judge by visual inspection of negatives, what filters should be used in printing.

The usual practice is to make a trial print at some predetermined printer setting, and then evaluate the print to determine corrections that may be needed.

Influence of Viewing Conditions

A basic requirement of any color process is that it should be capable of reproducing a neutral gray scale. The reproduction of white, gray and black areas in an object or scene may not be neutral by measurement, but as a rule the image should have a neutral appearance. There may be some situations, however, in which the neutrality of the gray scale has to be modified to some extent for the sake of pleasing picture appearance. Sometimes, too, color balance has to be mis-adjusted deliberately to achieve unusual pictorial effects.

The appearance of color pictures is affected strongly by the spectral energy distribution of the viewing illumination. Usually, films in the 35 mm size are projected with arc lamps. This light is quite blue in color compared with the light from a tungsten lamp normally used for 16 mm projection. These factors have to be taken into account in attempting to judge color balance or to modify the printing process.

Adjusting Color Balance

The magnitude of the silver images formed in the three layers of a color film during first development depends on the amounts of red, green and blue in the exposing light. The amounts of cyan, magenta and yellow dyes formed in the film layers are dependent in turn, on the magnitude of the silver images. When a color print is made, the amounts of red, green and blue light reaching the print film depends on the color of the illumination modified by the color absorbing and transmitting characteristics of the color negative (or reversal original) film.

If an original reversal film is too red, for example, the light reaching the print film has too much red in it, and the red-sensitive layer of the print film receives too much exposure. To compensate for this undesirable condition, the color of the printing light is modified by altering the relationship of the red to green and blue. In a subtractive printer this is done by placing a cyan filter in the printer light beam to absorb some of the red light. In an additive printer, where three colored lights are mixed to obtain the printer light, the amount of red light may be reduced, or alternatively the amounts of blue and green light may be increased, by adjusting the vanes in the appropriate light beams.

Color Timing (Grading)

Various methods are employed to estimate the kinds and degrees of adjustment required in printing to compensate for negatives (camera originals) that are too light or too dark, or that have some type of undesirable color balance. The procedure is known as timing, or grading, and includes the preparation of some form of corrective instructions for the printer. Instructions for manually controlled printing operations are in the form of a card showing the settings of the printer selector that the operator must make as the film is printed. The places in the negative (camera original) where these changes must be made may be indicated by notches punched in the edge of the film, actuating a light-changing mechanism. Alternatively, small patches of aluminum foil may be attached to the edges of the film to perform this operation electromechanically. In semi-automatic printers, timing instructions may be in the form of a perforated tape.

Traditionally, film timing has been performed by a motion picture technician who examines the camera originals over an illuminated surface and estimates the degree of correction required, scene-to-scene, to obtain prints that will have a satisfactory appearance when projected on a screen in the laboratory review room. This is a task that requires considerable skill and experience.

While the first, or trial, print is being screened, notes are made of the portions that would benefit from further corrections. The printing instructions are modified accordingly and a second print is then made incorporating these changes. As a rule, the purchaser or client has the right to attend the screening of the trial (answer) print and ask for changes to be made. If the purchaser does not exercise this right, prints are made at the discretion of the laboratory.

Timing Color Negatives

The corrections required to obtain acceptable prints from reversal originals can be estimated fairly accurately by a skilled timer visually examining the originals over an illuminated surface. Timing color negatives is much more difficult, because of the presence of colored couplers in the film and the reversal of colours in relation to those in the original scenes.

Several methods may be employed to determine the most favorable printer exposure conditions. The simplest is to make a series of trial prints, viewing the results of each trial, until acceptable results are obtained. Variations of this method involve determination of printing constants so as to minimize the number of trials.

In printing color negatives, a difference in overall exposure of about 0.04 log exposure unit (about one step in a subtractive printer exposure scale) is likely to be apparent in the print. An even smaller change in exposure of one layer relative to the other two is likely to be apparent as a color balance change. The variations in exposure level and color balance that can be tolerated in a print depend on the nature of the scene, the subject, its brightness, and whether or not the scene is to be reproduced naturally.

Scene Testing

The effect likely to be obtained from a given color balance change in the printer may be estimated roughly by viewing a test or trial print through viewing filters made up from combinations of CC filters. A better method is to make a series of exposures of selected scenes with a device such as a scene tester.

The basic elements in a scene tester are a light source, a platen for holding a series of CC filters, a light-changing mechanism and a means for exposing the film. This arrangement is intended to simulate the exposure conditions in a subtractive printer. The negative is placed on a pair of rewinds and threaded through the exposure mechanism with a length of unexposed print stock. When the exposure mechanism is actuated a complete series of exposures is made, light of a different color being provided for each frame in the series. A typical series is 16 frames. After an exposure test has been made on each scene in the film program, the print is processed and examined over an illuminated surface. From these test strips the frames with the most favorable picture appearance are selected, and related to the CC filters fitted in the corresponding exposure steps in the scene tester series.

Color Printer Calibration

Owing to the extreme sensitivity of color film to slight changes in the color of the exposure light, it is essential even in the simplest methods of printing, to maintain uniform spectral energy distribution in the printer illumination. This can be accomplished with a light-measuring device, such as a photo-electric cell and microammeter. The cell is mounted in a probe, for insertion in the printing aperture, and with the aid of red, green and blue filters, readings can be taken showing any changes in the color of the light source. By selecting filters with transmission characteristics similar to the spectral sensitivity characteristics of the color print film, a more accurate calibration of the light source can be achieved.

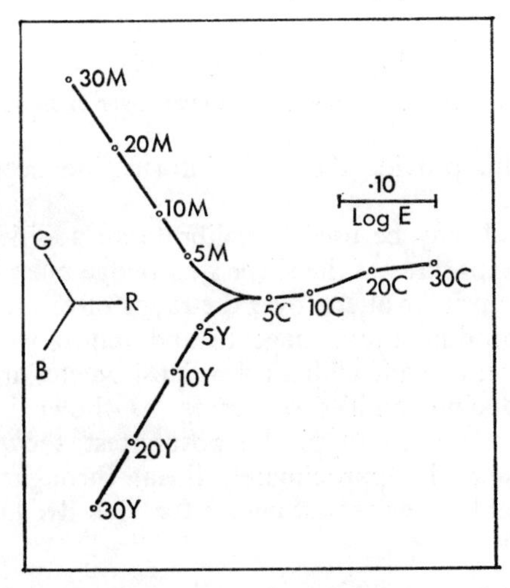

Fig. 55 Calibration of a subtractive color printer with a photo-electric meter.

There is another, equally important consideration, Without a knowledge of the shifts caused by changing CC filters and exposure level in subtractive printing, or by attenuating the red, green and blue light in additive printing, accurate color timing is almost impossible. Fig. 55 shows the results obtained with a good subtractive printer calibration method. Here the changes in exposure and color balance of a typical print film are compared with the corresponding meter readings expressed in log E units.

The production of properly balanced color prints with subtractive printers is probably better understood and somewhat more easily accomplished than additive printing due to the widespread use of scene testers. The relationship between film exposure and printer light changes may be established fairly easily. A film loop is prepared, including a calibrated silver step tablet. This scale is printed on color print film with a range of color compensating filters. Measurements made with a color densitometer on the

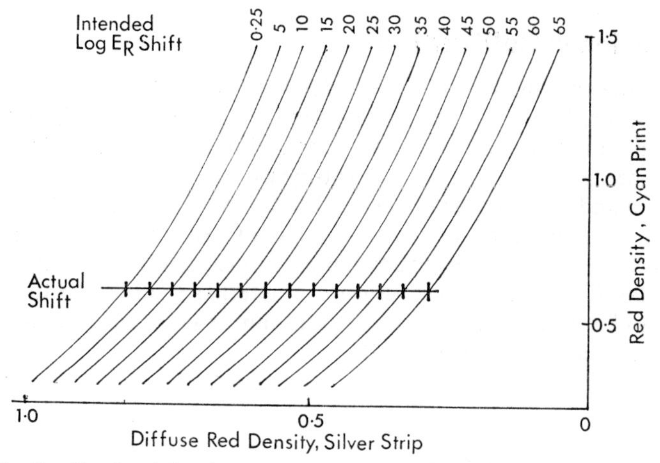

Fig. 56 Family of red density curves for the cyan layer in an additive printer test.

processed print film provide the basic information needed for accurate printer calibration.

A similar method may be used to calibrate an additive printer, but in this case separate exposures of the silver step wedge must be made with the red, green and blue printer filters, over the range of printer exposure levels. These exposures produce cyan, magenta and yellow wedges. Red, green and blue measurements made with a color densitometer are then plotted on graph paper, producing families of curves, as shown in Fig. 56. These curves, for the cyan layer in a typical exposure test, show that the scale of printer exposure steps is approximately linear throughout the range. A test of this kind yields three sets of curves for each dye layer, or a total of nine sets altogether.

Timing for Additive Printers

Additive printing offers the advantage of much better color print quality, but timing presents a major difficulty. The additive timer is faced with the problem of estimating the settings of three colored light beams required to produce a particular change in the appearance of the resulting color print.

Printer calibration methods, described above, can be used to show the effects of printer settings, but the timer has to translate this information into subjective terms—the effects of these changes on the appearance of color pictures.

To avoid the uncertainty of visual timing and to improve the yield of acceptable prints, non-subjective methods of color timing have been devised. There are several possible systems by which printing exposure conditions may be determined. If, with each negative to be printed, a gray scale chart has been photographed using the same lighting conditions as the actual scene, color densitometer analyses of the gray scale in the negative should furnish sufficient data for the computation of approximately correct printing conditions.

Another method is to make an analysis of the integrated density of a frame from each scene, in terms of red, green and blue densities. From these measurements, printing constants may be computed which, on the average, should yield satisfactory prints.

The first of these methods is slow and tedious, and while prints made in this manner may be said to be densitometrically accurate reproductions, the appearance of the resulting pictures may not be esthetically pleasing. While the second method is much simpler, it can be relied on only for average scenes, photographed in normal conditions.

Color Timing Calculators

A much more attractive method for calculating additive printer settings is to view the color pictures in a device that provides means for changing the color of the illumination. With such a system it is assumed that a change in the color of the illumination is approximately equivalent to the change in the color balance of the print being viewed. This method can be used to determine from a trial print the changes in printer settings needed to obtain best possible picture appearance in each scene.

In a typical instrument of this kind, changes in color of the viewing illumination are shown in terms of red, green and blue meter readings. By means of an appropriate calibration procedure, these meter readings may be transformed into corresponding changes in the settings of the red, green and blue light beams in an additive printer.

Recently, this concept has been extended to enable analyses to be made of color negatives. Devices for this purpose must be capable of reversing the colors in the negative and cancelling out the effects of the colored couplers on the appearance of the color pictures.

With the aid of a calibrated display device showing positive color pictures, the adjustments in negative reproducing conditions needed to obtain pleasing pictures may be translated into corresponding additive printer settings.

Electronic Analysis Methods

Equipment has been available for several years by which negative color pictures may be analyzed electronically. It is not difficult to set up an electronic reproducing condition simulating exposure conditions in an additive printer, so that adjustments in the electronic reproducer may be translated directly into red, green and blue light settings. Equipment of this type makes use of color television techniques to produce positive color pictures from color negatives. The pictures appearing on a color monitor tube may be compared side-by-side with a reference picture obtained by direct rear screen projection of a high quality positive color print*.

Automatic Printing

Most printing methods presently used are only semi-automatic, in that laboratory operators must evaluate films scene-by-scene, and modify the printer instructions in accordance with these visual evaluations. This human link has been a major obstacle in achieving fully automatic film printing systems. A more efficient system of production would be possible if the burden of operator judgment could be lightened by new methods of film printing. The result would be increased productivity, more uniform print quality, decreased color costs, reduction in operator skill requirements and simplification of inspection procedures.

Attempts to achieve complete automation of color printing systems have been unsuccessful in the past. Additional subjective evaluation has been found to be desirable to compensate for excessive or unusual variations in the printing materials, and to achieve best possible picture appearance in every conceivable situation.

There is another, much more promising, approach to these problems—that is, to set up a standard color film process, and deliberately expose the original camera films to conform with this reproducing condition. A clear distinction should be made between amateur and professional film operations. A laboratory providing printing and processing services for amateur movie makers must have a flexible system to accommodate the maximum range of variation that may be encountered in the camera originals. Professional film production, on the other hand, should be organized in an entirely different manner. A trial print made at the standard or normal settings of the printer would show the effects of any variations in exposure level and color balance of the camera originals. With the aid of a suitable color analyzer, the corrections needed to compensate for these errors could be determined and programmed into the printer for additional prints. This procedure would bring to light, first of all, the nature and extent of any variations in the camera originals, and at the same time, indicate the printer settings required to obtain best possible prints from the originals. Obviously, as the causes for variations in the camera originals are reduced or eliminated, the need for corrections in printing would be minimized, until eventually there would be no need for a second trial print.

* See *An Instantaneous Electronic Color Film Analyzer*, by Loughlin, Page, Bailey, Hirsch, Miller and Giarraputo, *SMPTE Journal*, Vol. 67, No. 1 (January 1958), pp. 17–26.

9 Sound on color film

The methods employed to record sound for programs on color film are much the same as those used for black-and-white film programs. As a rule, sound is recorded originally on a separate magnetic tape or film machine running in synchronism with the picture camera. Afterwards, a mixed magnetic track is made incorporating music, effects, etc., with the originally recorded dialogue. This master magnetic sound film may be played back in sync with the picture film when the program is being broadcast, or the sound may be transferred to the picture film in the form of an optical or magnetic track.

For some kinds of television film work—for example, news and sports—it is usually more convenient to make use of magnetic-striped camera original materials, so that both sound and pictures can be recorded simultaneously on the same film. If necessary, the sound may be dubbed off the camera original, to another magnetic film, during subsequent editing or program assembly operations. Sometimes, problems may be encountered when films with a magnetic stripe must be intercut with optically recorded tracks. To cope with situations of this kind, telecine projectors must be equipped with a magnetic playback head, in addition to the normal optical sound reproducer.

An optical sound track on film consists of a narrow area at the side of the picture frames in which the sound record appears in the form of continuously varying patterns. These patterns may consist of variations in density (variable density), or variations in the width of the track, composed of a clear central area with high density boundaries (variable area). These two types of optical sound track are shown in Fig. 57. The most popular method of optical sound recording is variable area, but both types of track may be encountered in television films.

Optical recordings are made by exposing the film with a light modulator which translates variations in electrical signals into density or area patterns on the film. As the film moves at a steady rate over a sound drum, a fine line of light focused at the plane of the film produces the sound track exposure.

To reproduce an optical track, the film is transported at the same rate over the sound drum in the projector, where a narrow beam of light focused on the film is modulated by the varying patterns in the film. A photocell or transducer located on the opposite side of the film generates varying output signals, and these signals, after being suitably amplified, are fed to a

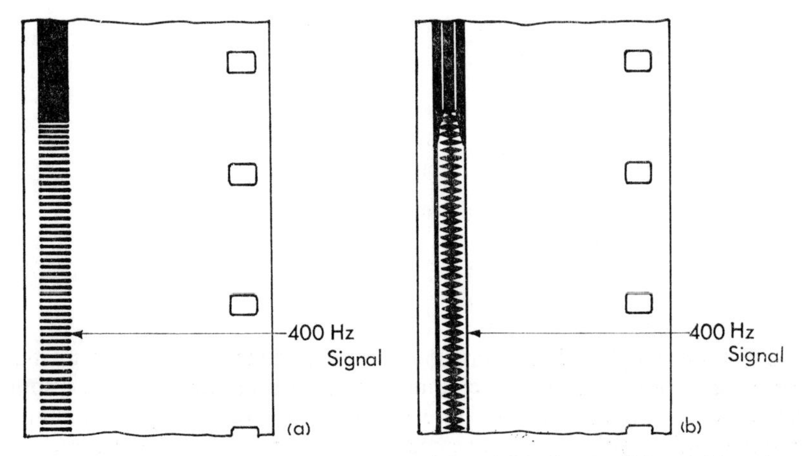

Fig. 57 Types of optical sound recording (a) variable density (b) variable area.

loudspeaker to reproduce the original sound. Ideally, the audio signals recovered at the output of the sound reproducer should be as nearly as possible identical with the signals applied to the modulator in the recorder. For various reasons, some of which will be detailed in following paragraphs, these ideal conditions are very difficult to achieve in practice. Sound recording on color film stock gives rise to special problems, especially in processing and reproducing stages.

Sound Track Characteristics

The techniques of optical sound recording were developed long before color films came into general use. Sound tracks on black-and-white films are processed in the same solutions and for the same lengths of time required for the accompanying picture images. The spectral density of a photographic silver deposit is such that a receptor with almost any response characteristic might be used in sound reproducers. Receptors with maximum response in the near-infrared region of the spectrum were selected to give high sensitivity with tungsten light sources, which radiate maximum energy in this region.

The introduction of color films complicated the processing problem because it is difficult to retain silver in the soundtrack area while removing it from the picture areas of the film. Dyes in color films have relatively little absorption in the infrared. To obtain sound tracks on color films that can be reproduced with phototubes having infrared sensitivity, the sound track area of the film and the picture areas must be treated differently in processing.

To avoid the need for special processing of the sound track area of the film, projectors could be fitted with receptors having sensitivity in the visible region of the spectrum. Lead sulfide phototubes, with S4 response, such as the RCA Type 934 vacuum phototube, can be obtained. However, it is

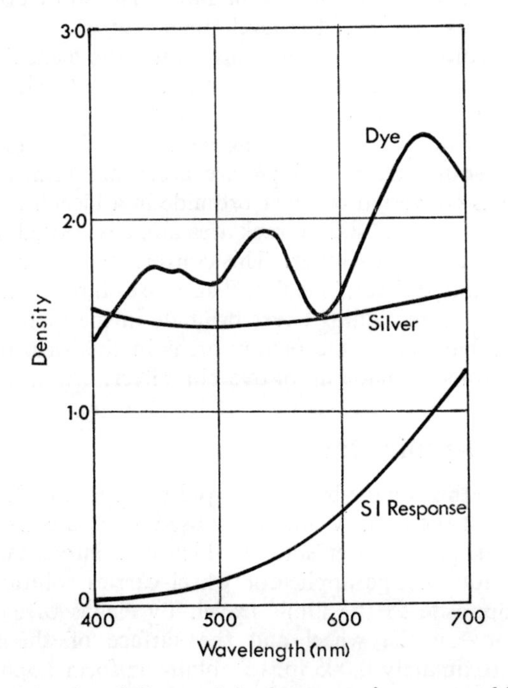

Fig. 58 Spectral density of silver and dye images and response of S1 photosensitive receptor.

unlikely that all projectors in use would be modified to reproduce dye sound tracks properly. In any event, the dyes in color films are selective absorbers, as shown in Fig. 58.

Selective absorbers introduce variations in sound reproduction with differences in the maximum spectral response region of receptors. To reduce the distortion resulting from a change in spectral response of the receptor requires the use of at least two dyes. Frequency response is affected by the position of the track layer in the film. The use of two dye layers would result in a difference in frequency response with phototubes having different spectral responses.

The less selective an absorber is, the less variation there is in sound reproduction characteristics. Silver deposits are relatively non-selective. As silver in any film layer contributes density at all wavelengths, the relative exposure of the film layers may be altered to obtain considerable control of the shape of the film's characteristic curve. The effect is likely to be similar for different types of receptors.

Dye and Silver Tracks

Several methods have been proposed for retaining or recovering the silver in the sound track areas of color print films. The most commonly used method is known as differential redevelopment, in which the silver in the sound track is reconverted to metallic silver after the bleaching stage.

During color print film processing, the sound track and picture images are developed simultaneously to dye and metallic silver in the color developer. In the next step of the process—a fixing bath—all of the unexposed silver halides in both sound track and picture areas are removed. Next, the developed silver is converted to silver bromide in a bleach bath.

Following the bleach, the sound track area alone is treated by the application of a reducing agent (developer). This converts the silver bromide in the sound track to metallic silver. The film then passes through a second fixing bath where all of the remaining silver bromide in the picture areas of the film is removed. This leaves the picture areas in the form of dye images, while the sound track is made up of dye plus silver.

Sound Track Applicator

Many different methods have been employed to apply the developer in the sound track area of the film. A commonly used method consists of a rotating applicator wheel, the lower side of which dips into a tank containing the developer solution. The applicator wheel carries solution to the film in a direction opposite to the film's travel. By means of a dial indicator the distance between the wheel and the surface of the film may be adjusted to approximately 0.005 in. to obtain uniform application of the solution.

The developer must be capable of uniform action without agitation, which requires the use of a powerful reducing agent, capable of converting all of the silver salts to metallic silver. A high reaction rate is essential to reduce the time of the reaction as much as possible. To minimize the tendency of the solution to spread beyond the area of the sound track, it should have a high viscosity.

Silver Sulfide Tracks

With some types of reversal color film, the sound track area is converted to silver sulfide by the application of a special solution. The process consists of three basic steps.

First, the exposed silver halides in the picture and sound track areas of the film are converted to negative silver images in the first developer. Next, the silver halides remaining in the sound track area are converted to silver sulfide by means of an applicator. In the following color developer, the remaining silver halides in the picture areas are converted to silver and dye images. Finally, the silver images in both sound and picture areas are removed with a bleach solution. This leaves positive dye picture images and a reversal silver sulfide sound track—that is, the portion of the sound track exposed during printing becomes the transparent area in the print.

The formation of silver sulfide in the three layers of the film can be controlled during exposure of the sound track, by varying the color of the printing light. The most favorable exposure conditions may be determined by

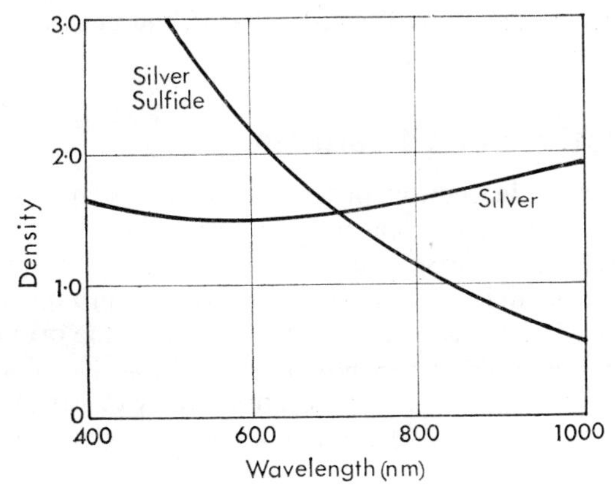

Fig. 59 Spectral densities of silver and silver sulfide sound tracks.

intermodulation tests for variable density tracks, and cross modulation tests for variable area tracks.

The color of the light used to expose the sound tracks controls the potential quality of sound reproduction which may be obtained with reversal color films. The attainment of a given density is not sufficient to ensure that the film has been properly exposed. Density is significant only when the color of the exposing light is rigidly maintained.

Silver Soundtracks on Reversal Color Films

Silver sulfide sound tracks with characteristics that are optimum for reproduction with receptors having S1 response are likely to be grossly in error for any other type of receptor. Both distortion and signal-to-noise ratio of silver sound tracks are somewhat less dependent on receptor response. Reproduction of color films in many projectors can be improved without impairing the result in any other projector by the use of silver rather than silver sulfide sound tracks.

The silver sound track process requires two separate solution applications in the sound track area only. First, the exposed silver halides in both picture and sound track areas are converted to silver in the first developer. Next, a fixing solution is applied to the sound track area, to remove all of the remaining silver halides. The third step is color development, during which silver and dye images are formed in the picture areas. Then, the silver images in both picture and sound track areas are treated in a bleach solution, which reconverts the metallic silver images to silver halide. In the next stage, the sound track image is re-developed to metallic silver. Finally, the film passes through a fixing solution, leaving dye picture images and a silver sound track. By applying a fixing solution in the sound track area only following first (negative) development, no dye is formed in this area of the film during color development.

An important consideration in this method of sound track processing is that negative silver images are formed—that is, the unexposed areas of the track are transparent in the print.

Sound Negatives and Positives

A common practice in motion picture printing operations is to make a sound negative from the original magnetic recording or magnetic master. The sound tracks on release prints are obtained by running the sound negative through a high-speed continuous contact printer with the print film, the exposure being restricted by a narrow slit in the printer aperture to the sound track area only of the film.

Sound negatives may have either variable density or variable area tracks. The use of variable area tracks is preferred, due to greater ease of controlling exposure and processing.

A positive variable area track has a central clear area, as shown in Fig. 57. In a negative track, the central area has high density, while the outer edges are clear. When a negative track is being printed, light reaches the print film through the clear areas, while light is prevented from reaching the print film by the central high density portion.

When positive color print films are processed, dye and silver tracks are obtained by a differential re-development procedure. The tracks obtained in this way are negative—that is, the exposed areas have high density, while the unexposed areas are clear. Thus, when a variable area sound negative is printed on positive color print film, positive sound track images are obtained.

Sound tracks on reversal color films processed to give silver images can be printed from sound negatives also. However, sound positives are needed for the printing of reversal color films which, after processing, have sulfide tracks.

When a transfer is made from magnetic sound recordings to motion picture film, either negative or direct positive images may be obtained by suitable adjustment of the film recording equipment. The most favorable exposure and processing conditions may be determined by means of a cross modulation test procedure.

Electrical Printing

Electrical printing consists of making sound transfers from the master magnetic film directly to the release prints by a recording procedure. The sound printer is fitted with a light modulator (galvanometer) which varies the exposure of the film as it passes a narrow illuminated slit, in response to the varying signals from the magnetic tape. This method of sound printing offers a number of advantages over continuous contact printing from sound negatives (or positives)—mainly better uniformity, freedom from wow and flutter, and improved frequency response. However, a separate sound printer is required for electrical printing, and the speed of printing is limited by the frequency response characteristics of the light modulator. Normally, electrical printing is carried out at standard sound speed—90 ft per minute for 35 mm and 36 ft per minute for 16 mm.

Distortion in Film Sound Tracks

Ideally, the boundary between the clear and dark portions of a variable area sound track should be sharply defined, with the clear areas having no density, and the dark areas being completely opaque. These ideal conditions cannot be achieved in practice, owing to the manner in which film images

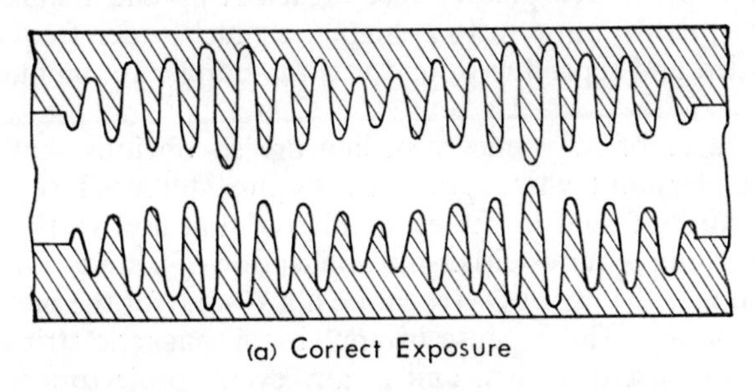

(a) Correct Exposure

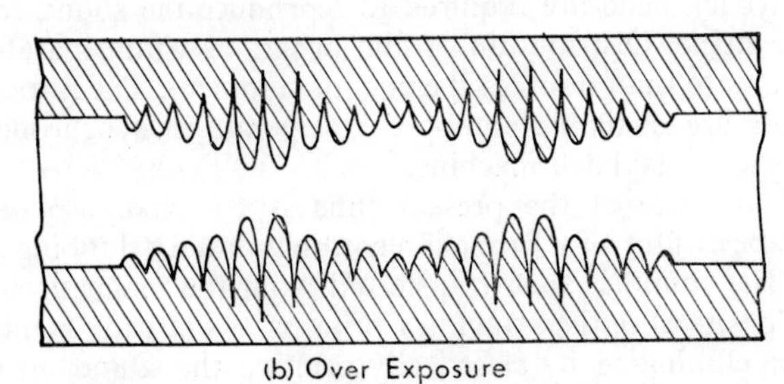

(b) Over Exposure

Fig. 60 Effects of image spreading in variable area sound tracks.

are formed. There is a tendency, too, for spreading of the images, the result of light scattering during exposure. This leads to filling in of the clear areas of the track, as shown in Fig. 60, causing partial rectification of the high frequencies.

By deliberately distorting the wave patterns in the sound negative, a pattern closer to the desired form can be obtained in prints. This can be done by suitable adjustment of the exposure levels when the negatives and prints are being made.

To determine the most favorable exposure conditions, amplitude-modulated high frequencies are recorded and printed over a range of exposure levels. By measuring the rectified component in the prints, the best combination of negative and print exposure levels can be ascertained. These exposure levels are then related to the corresponding track densities by measuring the tracks with a densitometer.

When sound is recorded directly on release prints by an electrical printing process, there is no opportunity to cancel image spreading effects. However, the electrical wave-forms applied to the light modulator may be pre-distorted in the desired direction to achieve compensation for cross-modulation.*

Optical vs. Magnetic Sound Tracks

In comparison with optical sound tracks, magnetic recordings are superior in several respects—mainly in thc far smaller number of variables to be controlled, and the much greater ease of achieving and maintaining high quality. The signal-to-noise ratio is higher, a much wider frequency range can be recorded and reproduced, and it is far simpler to maintain uniform signal level.

The advantages of magnetic recording are so obvious that it may be difficult to understand why optical tracks are still used so extensively. The reason can be found in the economics of film use. In the first place, magnetic recording requires either the application of a stripe on the edge of the picture film, or the use of two separate films—one for the picture and the other for the sound. The expense of applying a magnetic stripe adds considerably to the cost of prints, and in any event, projectors fitted with a magnetic playback head are required to reproduce the sound recorded on the magnetic stripe. Distribution of film programs with separate magnetic sound tracks more than doubles the cost of shipment. Then, too, relatively few projectors are available equipped to operate in synchronism with a separate magnetic playback machine.

The usual practice at the present time is to make use of magnetic materials—tape or film—for original recordings and for dubbing and mixing operations. The sound is then transferred from the magnetic master to a negative (or positive) optical track for printing. As already mentioned, one stage may be eliminated by electrically printing the sound on the release prints, but this is not a popular procedure, mainly because of the much slower speed of this transfer method.

For some purposes, as in the preparation of programs on film for internal use by television stations, it is usually advantageous to provide separate

* For further information on this subject see the March 1954 issue of *SMPTE Journal*, p. 77, *Cross Modulation Compensator* by Singer and McKie.

interlocked magnetic playback facilities in telecine. The master magnetic film can be used for the on-air release of the program, thus preserving optimum quality up to the final playback stage and the cost of making an

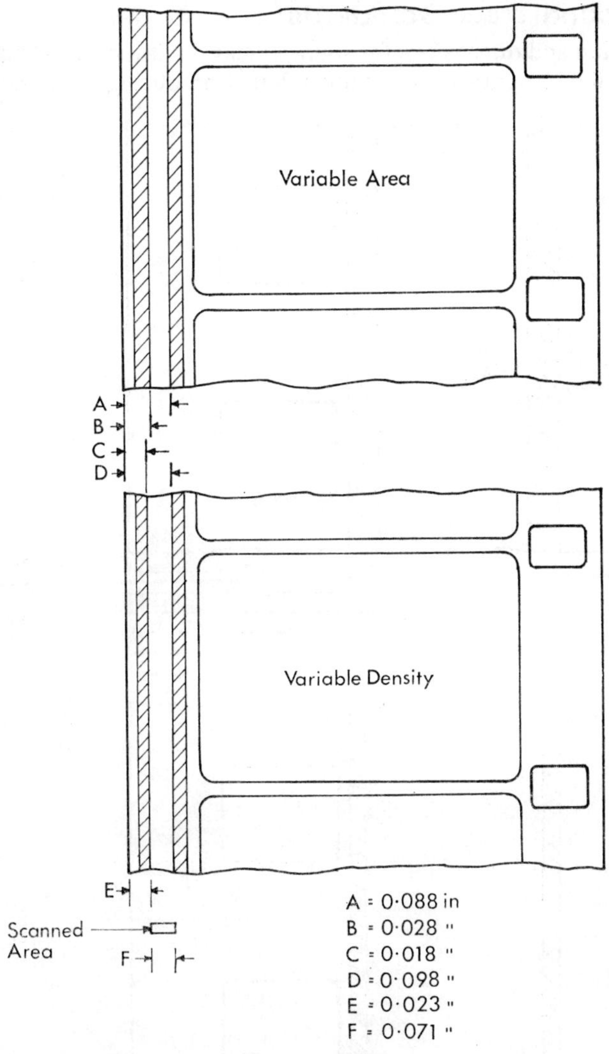

Fig. 61 Optical sound records and scanned area in 16 mm films.

optical sound negative (or positive) and printing the sound track on the picture film can be avoided.

Most television stations find that it is advantageous to equip at least one telecine projector with magnetic stripe playback facilities. This greatly simplifies the recording and reproduction of sound for news, sports, interviews and public affairs programs. Striped reversal color film provides program producers with a means for obtaining high quality pictures and sound on the original camera film in a single processing operation.

Film Sound Track Standards

The location and dimensions for sound tracks on film must be standardized to enable film programs from many different producing organizations to be

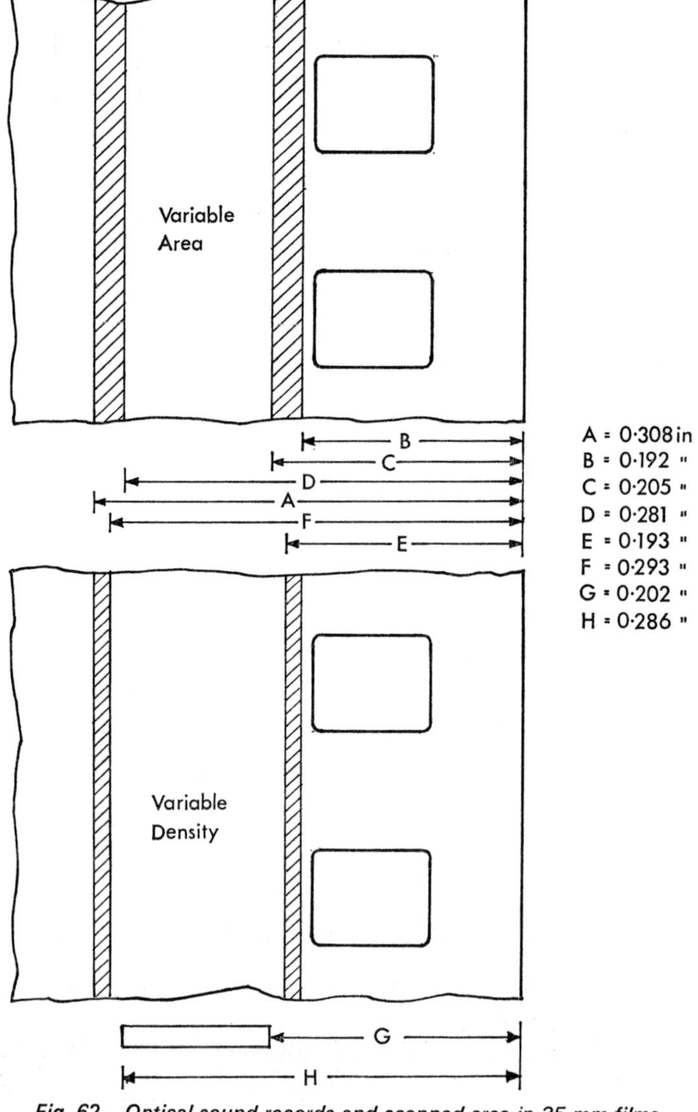

Fig. 62 Optical sound records and scanned area in 35 mm films.

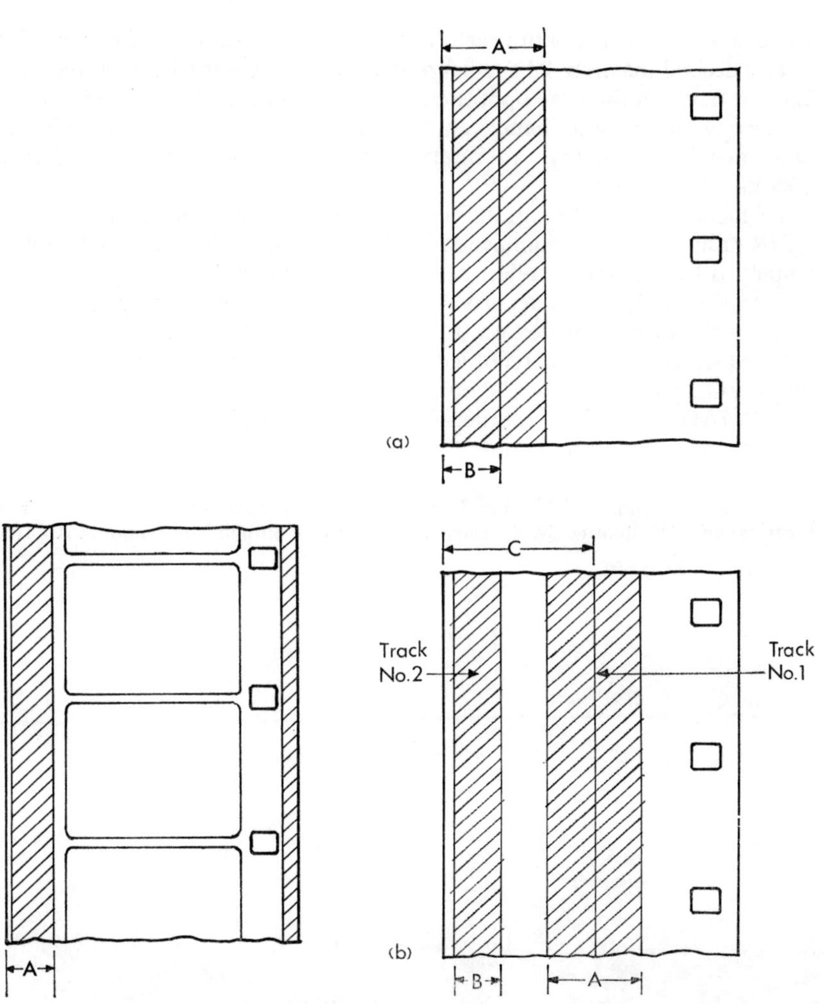

Fig. 63 (above)   Magnetic stripe on 16 mm film. A = 0·100 in. (100 mils).

Fig. 64 (right)   Magnetic film recording standards. (a) North American edge track: A = 0·200 in (200 mils), B = 0·103 in. (b) European center track. A = 0·200 in., B = 0·100 in., C = 0·315 in. (8 mm).

reproduced on any telecine projector. Fig. 61 shows the standard location and dimensions of 16 mm optical sound records and scanned areas in projectors. The corresponding 35 mm optical sound track standards are shown in Fig. 62.

Films with magnetic stripe in the 16 mm format have the oxide coating applied in the sound track area as shown in Fig. 63. A balancing stripe on the opposite side of the film outside of the perforations facilitates even winding of the film in large reels.

Perforated magnetic film for 16 mm recording usually has a uniform oxide coating over the entire surface, and the sound track could be recorded in any desired location. The usual practice in North America is to record a 200-mil track on the edge opposite the perforations, as shown in Fig. 64(a). In Europe, on the other hand, center track magnetic recording is preferred, as shown in Fig. 64(b). Alternatively, the sound may be recorded in a 100-mil edge track.*

To facilitate the international exchange of television programs on film, CCIR designates, in Recommendation 265, the various formats that may be employed by a series of code words. Following are a few examples—

COMOPT—Sound is recorded as a variable density or variable area track in the space provided on the edge of the picture film.

COMMAG—Sound is recorded on a magnetic stripe in the area of the film normally occupied by the optical sound track.

SEPMAG—Sound is recorded on a separate perforated magnetic film, synchronously with the picture film.

* See Recommendation 265 *Standards for the International Exchange of Television Programs on Film* published by the International Radio Consultative Committee (CCIR).

10 Color slides for television

One of the most useful, versatile and inexpensive methods for reproducing color pictures in the television system is the 2 × 2 in. color slide. Slides are made by photographing scenes or specially prepared art work in the form of double-frame exposures on 35 mm color film. After processing, these frames are placed in 2 × 2 in. slide mounts ready for use.

The usual procedure is to make use of a 35 mm still camera, which exposes the film lengthwise in frames about 1 11/32 by 29/32 in. (36 × 24 mm) in size. These frames occupy a space on the film 8 perforations in length or about twice the size of the standard 35 mm motion picture frame. Film for use in cameras of this type is available in 20- or 36-exposure cassettes, or in bulk rolls that may be cut into suitable lengths for loading into cassettes.

The slide mounts commonly used in amateur photography consist of a cardboard folder with an opening cut out slightly smaller than the frame on the film. After the individual frames in a roll of film have been cut apart, a frame is placed in a folder, centered over the opening, and the two halves of the folder are then cemented together to hold the frame in place. Alternatively, the film frames may be placed in metal or plastic mounts fitted with cover glasses to hold the film in a flat plane during projection. These mounts are usually made slightly smaller than 2 in. square to avoid binding in inexpensive slide projectors that may not be accurately manufactured.

When slides made in the ordinary way for amateur photography are used in television programming, precise framing of the opening in the slides and location of the picture information within the frames relative to the television raster is very difficult if not impossible to achieve.

Early models of television slide projectors were provided with magazine-type holders, the slides being held in place by spring clips. Framing adjustments were made manually by an operator, the slide mount being shifted to one side or the other under the spring clips while the position of the picture relative to the edges of the raster was observed on a nearby television picture monitor.

Slide Registration

Significant improvements in the use of slides in television had to await the development of heavy-duty projectors, capable of registering images automatically. The basic requirements for accurate registration are:

1. Properly machined slots in the slide projector magazines, and precise positioning of the apertures in the optical system of the projector.
2. Full-sized slide mounts giving a snug fit in the slots of the magazine.
3. A means for accurately locating the film frames in the slide mounts.
4. Camera equipment calibrated to position picture information properly within the film frames during the original photography.

Precision Slide Mounts

In 1966, a recommended practice was published by the Society of Motion Picture and Television Engineers providing guide lines for the manufacturers of slide mounts for television use. Fig. 65 gives the dimensions of the

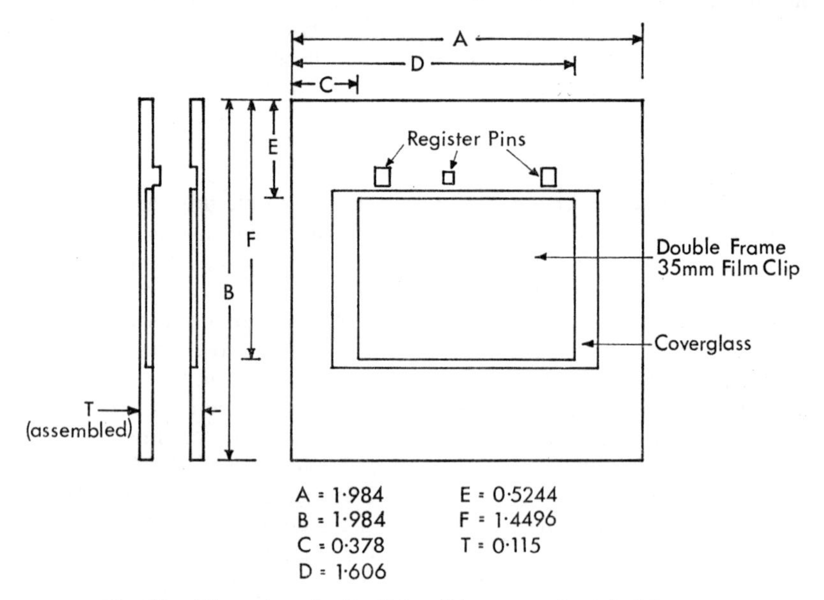

Fig. 65 Dimensions for 2 × 2 in. slide mounts for television use.

slide mount. The minus tolerance on the outside dimensions is only 0.005 in. to ensure a snug fit in the projector magazine. Register pins are provided in the mount to engage the film perforations, and thus accurately locate the position of the film frames relative to the opening in the mount.

This recommended practice also provides guidance for makers of slides for television use, to ensure that letters and other geometrical information is not tilted, and to permit lap dissolves to be made between slides with related subject matter.

The dimensions and tolerances in this recommended practice are based on the need to register information in successive slides within plus or minus five television lines*.

Precise Registration of Picture Information

To locate picture information accurately relative to the film perforations, a good quality camera is required. When the camera is used to photograph flat art work, it must be rigidly mounted relative to the surface in which the art work is placed, and some means must be provided to position the art work accurately relative to the aperture in the camera.

If the camera can be fitted with a ground glass or some other type of direct viewing device, it is a simple matter to place the art work in the desired location under the camera. However, few 35 mm cameras have this facility, and an indirect locating method must be employed. One fairly simple method is to make up a chart with ruled cross lines, each vertical and horizontal line in the ruling being numbered, beginning at the center. This chart is then photographed with the camera, and by means of successive test exposures, its position is adjusted to place the crossed lines at the center of the chart in the exact center of the film frame.

To accommodate art work of different sizes, a field chart may be made up, with numbered rectangles corresponding to fixed positions of the camera lens relative to the art work, as indicated on a calibrated distance scale. A much better plan is to arrange for all art work for television slides to be prepared on standard size paper board. Indexing marks placed on the art work by the artist may then be used to locate the picture information properly in the film frames.

Safe Title Area for Television Transmission

A very important factor in the production of television slides is the need to confine important picture information, such as lettering and commercial products, within the area of the raster that will be reproduced on the majority of home receivers. Fig. 66 shows the safe title area as recommended by the Society of Motion Picture and Television Engineers.†

If the telecine slide projection equipment permits one slide to be supered

† *SMPTE Recommended Practice RP9*, Dimensions of Double Frame 35 mm 2 × 2 Slides for Precise Applications in Television, *SMPTE Journal*, Vol. 75, No. 8, August 1966, page 755.

* *SMPTE Recommended Practice RP8*, Safe action and safe title areas for TV transmission, *SMPTE Journal*, Vol. 76, No. 6, June 1967, p. 572.

over another, a test slide showing the safe title area in the form of a black line on a clear background may be used to check quickly the location of titles and commercial products in slides—and films as well.

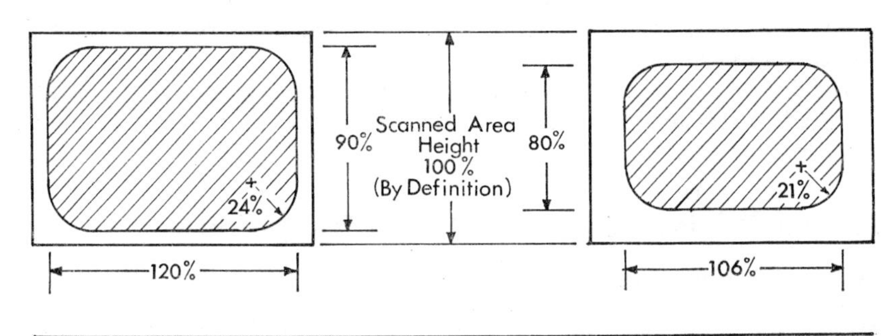

Dimensions of Safe Action Area			Medium	Scanned	Dimensions of Safe Title Area		
Width 120%	Height 90%	Radius 24%		Height 100%	Width 106%	Height 80%	Radius 21%
0·331 in.	0·248 in.	0·066 in.	16mm Film	0·276 in.	0·294 in.	0·221 in.	0·059 in.
0·713 in.	0·535 in.	0·143 in.	35mm Film	0·594 in.	0·634 in.	0·475 in.	0·127 in.
1·013 in.	0·759 in.	0·203 in.	2"×2" Slide	0·844 in.	0·900 in.	0·675 in.	0·180 in.

Fig. 66 Portions of the television scanned area within which titles and other important picture information should be located to avoid being cut off by the edges of the mask in home receivers (SMPTE RP8 – 1967)

Slide Production Techniques

Television advertisers are extensive and usually very effective users of slides, and most television stations provide facilities for telecasting considerable numbers of slides in rapid succession. One of the most annoying features of telecine video operation is the large and sudden change in video levels that so often occurs as slides are being changed. To compensate manually for these changes requires an alert, competent video operator. Video signal level adjustment can be effected much more quickly by automatic control devices, but signal level adjustments made in this way usually degrade and distort the picture information.

These signal level changes are caused by differences in maximum and minimum densities between successive slides. Density variations are inevitable when slides are made in the conventional manner—that is, by attempting to produce visually acceptable pictures from artwork as supplied. To compensate for different kinds of artwork, the photographer is obliged to modify film exposure conditions (and often processing conditions as well). Variations of this kind can be avoided by setting up a standard slide-making process, and insisting that artwork must be prepared properly for television reproduction.

Color Slide Production Control

It is a relatively simple undertaking to set up a control procedure for color slides that will yield uniform video signal levels, and at the same time ensure that the artist's conception is translated through the television reproducing system without change. It also ensures best pictute quality.

The basic reference is a standard slide reproducing condition in telecine as described in Chap. 14. The second requirement is a standardized film process, in which film exposure and processing conditions are reproducible within narrow tolerances. Third, a method of artwork preparation must be worked out that will produce the desired picture characteristics—and colors—when the slides are reproduced in the television system.

Setting Up the Film Process for Color Slide Production

Uniform, reproducible processing conditions for short lengths of color slide film are quite easy to set up and maintain. With these conditions established, the exposure level required to produce a given minimum density in the film for a particular type of art work and lighting level at the camera can be determined in a simple series of test exposures. Peak white density may be established with a particular type of paper board used by artists. With the aid of a photoelectric exposure meter, a suitable lighting level at the art work may be selected. The minimum density representing the lightest area of the art work—for example, the paper board—should be between 0.30 and 0.40 to avoid compression of the lighter area of the pictures, in the toe of the film's characteristic curve. (It is assumed that reversal-type color film is used to make the slides.)

Once the conditions for achieving the specified minimum density have been established, it is a relatively simple matter to obtain the desired pictorial results. Artists can be given complete freedom, with the procedure described here, in the choice of colors and the arrangement of visual materials, providing that the art work has some areas that will be reproduced at the minimum density level, and that some maximum black areas are provided to establish television black level.

Telecine Evaluation of Slide Production Procedures

When a slide production facility is being set up, and at frequent intervals thereafter, arrangements should be made to evaluate slides, taken from current production, on a standard telecine. In this way, any variations observed in signal levels may be related directly to errors in the slide-making procedure, or in the preparation of the art work. At the same time, the reproduction of the various colors used in the preparation of the art-work may be assessed, to determine any changes that should be made to achieve the desired pictorial or esthetic effects in the television reproduction. As outlined in Chap. 1, the appearance of a color pigment to the eye may be altered considerably when it is reproduced, first by the film process, and then as the film is converted into electrical signals in the television system.

11 Producing television programs on color film

Programs seen on television may be divided into four main source categories as follows:

1. Programs produced with television cameras and broadcast directly (live) to the public.
2. Video tape recordings of television camera pickups.
3. Films made especially for television—the telefilm serials seen regularly every week.
4. Films made originally for projection on motion picture screens, and adapted for television—old movies, documentaries and the like.

These different types of program are made by production groups with entirely different objectives. Live television programs depend for their popularity on the ability to give the public a grandstand seat as events are taking place. Program production with video tape follows to some extent established motion picture practices, in that scenes recorded at different times can be brought together and related by an editing procedure. However, at the time the original recordings are made on video tape, the producer can take advantage of the highly flexible and versatile television techniques developed initially in live broadcasting.

Motion Picture and Television Production Methods

The production groups involved in making film programs for television are motion picture oriented, and regularly turn to motion picture production when the opportunity arises. The techniques developed in many years of highly successful motion picture making are applied with only minor modifications. Many motion picture groups are still making films with direct screen projection as their primary objective even when it is known in advance that the films will be shown eventually on television screens.

Many programs now being made with live television cameras and video tape could be produced more easily and efficiently, and at less cost, on color film. On the other hand, most programs now being produced on film by motion picture groups could be turned out more easily and quickly, and at far less cost, by adopting television production techniques. It is in these areas that the greatest advances in television program production can be made.

Program producers who have become involved in television favor the use of television cameras and video tape, mainly because the technical crews assigned to work with them are so well trained and organized. Moreover, all of the work, from the original camera pickups in the studio or on location, to the editing of the video tape and the final release of the program on air, is handled by television operating groups with much the same background, outlook and objectives. An important consideration for the television producer is the ability to *see* for himself what is going on—he feels he has full control of his production at all times.

A major advantage of television is the much higher rate of program production that can be achieved, compared with conventional motion picture filming procedures. A television producer can expect to complete at least a half-hour of program time for each working day in the studio. The simpler forms of program can be turned out at an even higher rate. While more sophisticated production methods using video tape have had a somewhat adverse effect on productivity, television's potential for rapid, dynamic, low-cost production is available for anybody who wishes to take advantage of it.

Often it is said that the main attraction of video tape is the ability to replay the tape immediately after each take. This advantage is far outweighed by the opportunities film offers for viewing and editing on simple, inexpensive equipment, at the convenience of program production groups.

The Camera and Original Recording

There is no good reason why television program production on film should be slower and more costly than video tape. In fact, film equipment is much simpler and far less cumbersome than television; no set up time is required prior to recording, and the film camera offers the unique advantage of being a combination camera-recorder, with the recording material in a magazine attached to the camera. Only a power cord is needed to operate the camera—on location a battery pack may be used to provide power. Separate synchronous sound recording related to the film perforations is normal practice, enabling the producer to build up the program sound track from multiple sound sources, back at home base, after the original recordings have been completed.

Television offers film makers well-nigh unlimited opportunities for program production. At the same time, film offers television producers an exciting and versatile medium of expression, one that is seldom used effectively.

The greatest disadvantage of film in television today is the uncertainty of the results. It is more than a disappointment for a producer to find, when his program is broadcast, that the color pictures are quite different from what he expected. When this happens he is likely to acquire a marked distaste for film and turn to the use of more reliable video tape whenever the opportunity arises.

A producer, using film as a recording medium, finds that he cannot see what is being recorded; he must rely on the cameraman to select appropriate angles of view, and the most effective scene compositions. Looking through the tiny camera viewfinder doesn't help very much—the producer can only try to visualize the kind of pictures that are being impressed on the film.

The lack of easy mobility of the film camera is a severe handicap for a producer with television experience. The usual procedure in motion picture production is to set up the camera in a fixed position for long, medium and close-up shots, with the action being repeated after the camera—and often lighting as well—has been repositioned. When dolly shots have to be made, the usual practice is to lay down special flooring to avoid unsteady camera movement. The distance over which the camera has to travel is measured, and the lens is adjusted manually to maintain sharp focus as the camera is moved. All this is slow, tedious work, compared with television studio operations, where multiple cameras are normally used; smooth floors are provided for continuous camera movement, and the cameraman maintains sharp focus with a lever at the rear of the camera, as he observes the pictures in his viewfinder.

Control of Scene Contrast

Then there is the problem of exposure control. Anyone who has tried to make pictures with film knows only too well that a great deal depends on exposure. With color film especially, even slight errors in exposure degrade picture quality appreciably. Determining the correct exposure for each scene with a photo-electric exposure meter is an uncertain, time-consuming procedure. No doubt as the result of the phenomenal success of amateur photographers in making pictures with available light, there is a widespread belief, especially in 16 mm film work, that the use of artificial lighting is an expensive luxury, and worse still, that pictures with a 'natural' appearance can be obtained only by filming real life in its natural state—that is, with available light. Film makers are supported in these beliefs by the excellent results regularly achieved by skilled motion picture technicians in the laboratory. The normal practice in this kind of work is to adjust printing conditions—and often processing conditions as well—to obtain the best possible picture quality from the original materials sent in to them by clients and customers. This, in fact, is an essential part of the art and craft of motion picture making.

While it is possible to obtain pleasing pictures for direct projection with these methods, television reproduction of films made in this way is often disappointing. As outlined in Chap. 5, conventional motion picture lighting and exposure techniques result inevitably in non-uniform maximum and minimum image densities, and variations in video signal levels when the film is reproduced in telecine. These undesirable effects cannot be corrected in the laboratory when prints are made. There is only one way to avoid these variations—that is, to adjust lighting and exposure conditions at the time of the exposure of the original film in the camera with this objective in view.

The control techniques needed to ensure uniform minimum and maximum picture densities are quite simple. A knowledge of the behavior of the television system is helpful in gaining an understanding of these techniques and their application. The first and most important requirement is the presence in each scene of black and white picture references.

The purpose of these reference areas is to establish maximum and minimum signal levels in telecine. These should be the lightest and darkest areas in a scene, ignoring specular reflections and areas of shadow in which no detail is desired. Most scenes contain picture elements that can be designated as reference areas. However, if a light or dark reference area is not present, an object should be placed in the scene to provide the reference. In an interview, for example, the lightest part of the scene may be a lamp shade, and this area may be designated as the white reference. If there is no dark area in the scene that can be designated as the black reference, a telephone placed on the desk may be used for the purpose.

The next requirement is a means for measuring the brightness (luminance) of these reference areas. This can be done with a spot brightness meter, as described in Chap. 5. A measurement of the white reference area indicates the camera lens setting required to locate this part of the scene at the desired point on the film's characteristic curve. A measurement of the black reference area, compared with the lamp shade, shows whether the scene has too much or too little contrast for television reproduction. These are very simple measurements to make, requiring only a few moments, and they give the cameraman information about each scene that cannot be obtained with a conventional photo-electric exposure meter.

The term 'black and white references' should not be taken literally. These are the darkest and lightest areas in a scene which, after being recorded on the film, produce black level and peak white signal voltages in telecine reproduction. In a scene these areas may be colored, but a properly calibrated spot meter quickly identifies them.

Exposure Control with an Electronic Viewfinder

A far better and much more convenient method for setting up scene lighting and exposure control is to equip the film camera with an electronic viewfinder. A small vidicon or Plumbicon tube mounted in a suitable enclosure may be attached to the side of the camera, replacing or supplementing the conventional optical viewfinder. The output of the tube may be displayed on a small cathode ray tube similar to the viewfinder in a television camera. This gives a much larger and brighter picture than the optical viewfinder, but the sharpness of the picture is limited by the resolution of the television display. Some experimental equipment has been built in which the light path in the viewfinder is split, to provide an optical image for direct viewing and a video output signal for exposure control and monitoring purposes.

The video signal generated by an electron tube looking into the camera viewfinder can be displayed on an oscilloscope in the form of video waveforms, and used to control scene lighting and film exposure. As the scanning beam sweeps across the face of the tube where the viewfinder image is formed, the video output signal rises and falls in amplitude in relation to the intensity of light in the optical image. A bright area of sky produces a signal with very high amplitude, while the signal falls to a low level—perhaps to zero—in a dark shadow area.

Calibration of the oscilloscope is accomplished with a reference signal producing a display with maximum amplitude, relative to an engraved scale. When the video signal from the electronic viewfinder is applied to the oscilloscope, the resulting waveforms may be related to the graticule scale, giving an indication of the amplitude of the signal in comparison with the calibrating signal. Plate 5 shows a typical video waveform displayed on an oscilloscope, with a bright sky area shown by the arrow at 100 units on the graticule. Zero on this scale represents minimum signal level in shadow areas of the scene.

The next step in the calibration of an electronic viewfinder is to relate video signal amplitude to the level of exposure of the film in the camera. Here again, the presence of a reference white area in the scene is essential. In an outdoor scene this may be an area of bright sky. From the characteristic curves plotted in Fig. 42, the density for this reference white area may be determined. Now, all that has to be done is to make a series of trial exposures with the type of film to be used, to relate waveform amplitude to film density. By adjusting the gain in the control unit for the electronic viewfinder tube, signal level for the reference white area may be made to coincide with the line on the graticule indicating 100 units, when the light entering the camera is sufficient to produce the specified peak white density in the film.

With an electronic viewfinder attachment on a film camera it is no longer necessary to make light measurements in scenes to determine the correct exposure for the film. All that the cameraman has to do is adjust the camera lens aperture until maximum signal amplitude of 100 units is obtained, representing the reference white area in the scene. When this simple adjustment has been made, the film may be exposed with the assurance that the reference white area in every scene is always recorded at the correct density in the film.

If the contrast range of an outdoor scene is excessive, the waveform monitor shows it as compression in the shadow areas. By means of artificial lighting, signal amplitude in shadow areas can be raised to reduce or eliminate the compression. This, of course, also raises the density of these areas in the film. Similarly, faces and other areas of primary picture interest may be reproduced at any desired level on the waveform scale by the use of supplementary lighting.

Indoors, the waveform display provided by the electronic viewfinder may be used to raise or lower the illumination in different parts of the scene, and for the setting of the aperture of the camera lens, to provide correct exposure of the film.

Electronic Viewfinder as an Aid in Program Production

The video signals generated by an electron tube looking into the camera viewfinder can be used for other purposes besides lighting and exposure control. They may be displayed on picture monitors as well. Instead of trying to peer into the optical viewfinder on a camera, the producer can have a large picture monitor, placed at any convenient location, to observe the action as it is recorded on the film.

The ability to see what is being recorded on the film gives the producer the opportunity to provide more effective direction for the cameraman. With the aid of a simple intercom circuit, directions for camera positioning and picture framing may be relayed from the producer to the cameraman via a small cable.

An electronic viewfinder, providing large pictures for monitoring purposes, permits a film camera to be used in much the same way as a television camera. Instead of a fixed tripod mounting for the camera, a movable dolly may be used to give the camera much greater freedom and flexibility.

Multicamera Techniques

It is customary to equip large television studios with three or four cameras. The techniques of television camera use were developed in the days before video tape, when all programs were transmitted directly to the viewing audience as the action was taking place. While the video signals from one camera were transmitted, the other cameras could be moving to new positions for following shots. The output of each camera appears on a picture monitor in the control room, and by means of an intercom circuit the cameraman receives directions from the producer. A video switcher-mixer unit allows the output of any camera to be selected for transmission, and effects such as fades and mixes between camera outputs can be introduced.

Recording on video tape allows the recording to be stopped when it is necessary to reposition cameras. Surprisingly, this has not led to the use of single cameras, as in normal motion picture work. Most television producers find that it is easier and more convenient to record an entire scene on tape, using three or four cameras, and switching from one camera to another as required for long, medium and close-up shots, to achieve continuity of action. This is also one of the very best ways to increase productivity in the studio.

Equipment is now available for producing programs on film in a similar manner. Most of the work so far has been done with two or three cameras.

Sometimes, the cameras are run simultaneously; in other cases, the cameras are started and stopped as required to conserve film stock. The producer directs the operation from a control room where the pictures from the electronic viewfinders on the cameras appear on television monitors. The camera to be used for each shot is selected by a pushbutton which lights a lamp to expose a cue mark on the edge of the film in the camera. The program sound is recorded on perforated magnetic film in a separate synchronously running machine (see Fig. 67).

Camera Monitoring Methods

A simple vidicon or Plumbicon tube looking into the viewfinder of a film camera provides only a monochrome picture for monitoring purposes. This may be a handicap for some program producers. In the present state of development, even the smallest color television camera is much too large for use as an electronic viewfinder for a film camera. Eventually, no doubt, this type of equipment will be reduced considerably in size and weight, enabling cameras to be constructed with full color monitoring facilities. For the present, a monochrome electronic viewfinder provides a very effective means for controlling lighting and exposure, and a partial solution for the problems of camera control while recording is taking place.

Laboratory Control Procedures

When an electronic viewfinder is used to establish correct exposure of the film in the camera, processing conditions for the film must be very carefully controlled within pre-determined tolerances. Any variations in processing conditions will upset the calibration of the camera—the relationship between video signal amplitude and the densities in the film. At the same time, any processing variations may adversely affect the picture color balance.

Setting up exposure level for printing is quite simple when the camera originals have been properly exposed, and each scene contains black and white reference areas. The printer exposure level is adjusted to obtain a specified minimum print density from the density in the camera original representing the reference white area in the scene. A minimum density of 0.30 to 0.40 is generally considered to be optimum for the reference white area in color television films.

A so-called 'one-light' print should be made from the camera originals—that is, the same printer exposure level should be used for all scenes, regardless of their composition or content. When the print is reproduced in a standard telecine the waveform display will show any variations in image densities as changes in signal amplitude, and these variations may be directly related to errors in the exposure of the original camera film. Fault reports prepared during these evaluation sessions may be used to track down the causes of the variations.

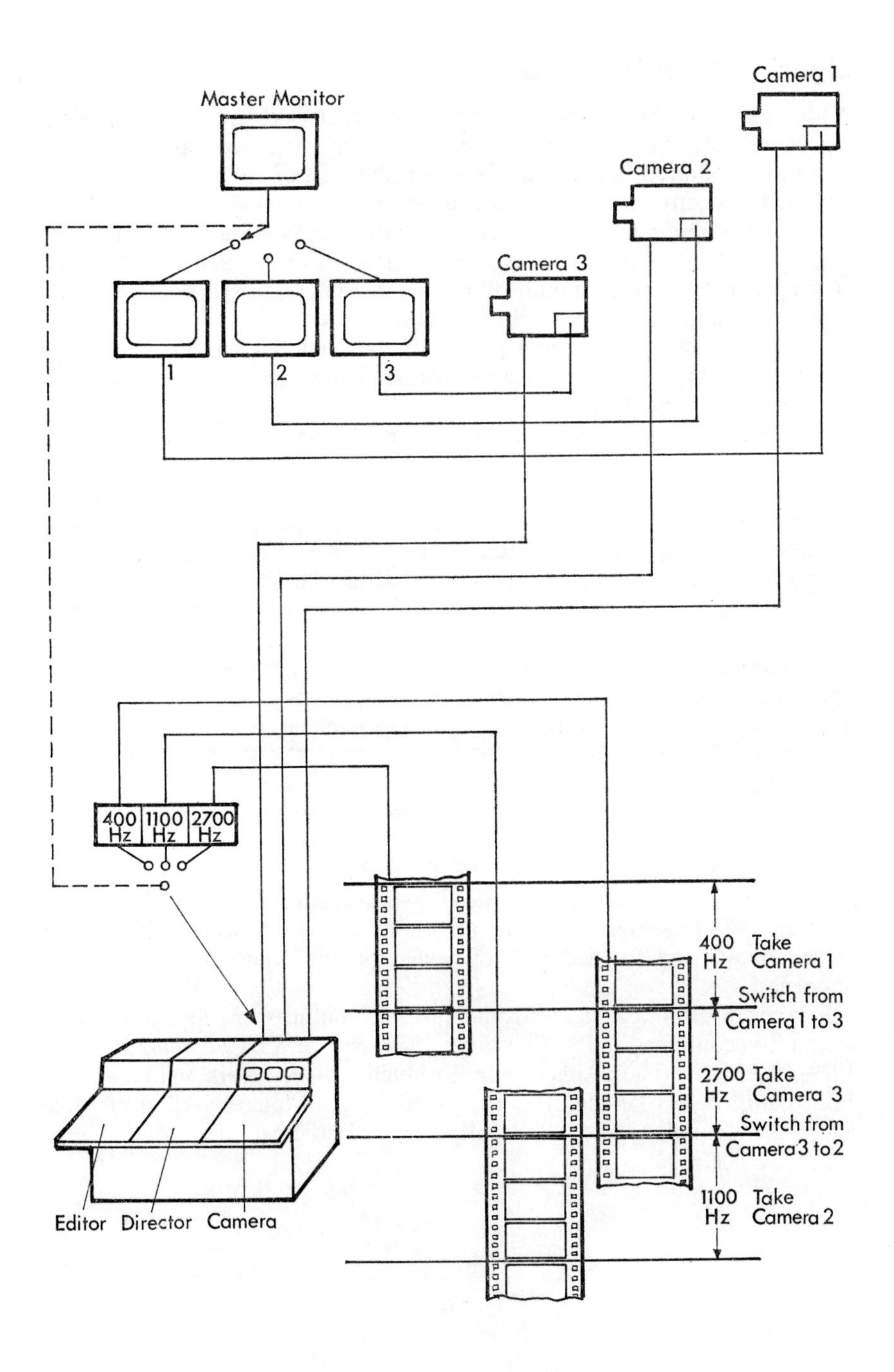

Fig. 67 Schematic diagram of a multicamera film installation.

Lighting Requirements

A film operation that has been set up in the manner described in the preceding paragraphs closely resembles a video tape recording operation. The equipment required to make the film recordings is, of course, much simpler and far less costly than television equipment. The quality of the pictures—and sound—obtainable with color film should be as good as, and perhaps even better than, video tape. However, successful film recording for television programs requires a quite different approach on the part of the camera crew, compared with motion picture making.

Television viewers see these programs on small screens in their living rooms. For this type of viewing situation, lively, dynamic camera work gives a program much greater interest. Television viewing does not require the niceties of motion picture production for large screens in darkened theatres.

Motion picture crews usually follow the practice of readjusting the lighting for long, medium and close-up shots. Much of the lighting equipment may be mounted on floor stands or attached to the top of sets. Setting up and readjusting lighting in situations of this kind is a slow, tedious process, that takes up a great deal of time while the crew and performers are idle.

Mainly because cameras in a television studio must be free to move in any direction, an entirely different approach to set lighting has been developed. Fixtures are suspended from an array of girders above the studio floor. The complement of lighting fixtures in a television studio is normally quite large. From a floor plan for the program, the studio crew can set up the lighting in advance, before scenery is put in place. In large, busy studios, lighting arrangements can be programmed during rehearsal into memory units.

Much of the lighting in television is far from ideal from a motion picture standpoint. However, television itself is a transitory medium, and most programs are seen only once—or at the most a few times. Little is to be gained by striving for a degree of perfection that viewers are unable to appreciate.

Further information on electronic aids and multicamera film production methods is contained in the December 1969 issue of *SMPTE Journal*, pages 1079–1082, "Television Aids to Film Production—New Version in a Mobile Film Unit," by Mike Metcalfe and Geoff Pryke, Molec Division of Mole Richardson (England) Ltd., Wembley Park, Middlesex.

12 Program assembly methods

Program assembly on video tape is a very attractive method of television production. The main advantage of video tape editing is the opportunity offered for rehearsing edits in advance, as many times as necessary, prior to making the actual splice or join between two scenes. Besides, after an edit has been made, the producer can redo the work if he wishes, by going back to the original recordings.

In comparison with video tape, conventional film editing procedures are slow, tedious and cumbersome. The usual practice is to prepare a work print first, by splicing together sections of prints from the camera originals. A list is made of the footage numbers appearing in the edges of the work prints and these numbers are used to locate the corresponding sections of the camera originals. Then the originals are cut and spliced together to match the work print, scene-by-scene. Finally the required number of prints are made from the edited camera originals.

As a rule, a good deal of extra footage is exposed during the original shooting, to give the film editor an opportunity to select the best portions of scenes and to lengthen or shorten shots as the nature of the program demands. In the preparation of the work print, films are viewed in an optical device which produces postcard size pictures. The work print also can be projected on a large screen in a review room to give the producer an opportunity to evaluate the progress of the work.

The conventional methods of film editing have been used with little change for 50 years or more. This is a highly individualistic type of work. A film editor, working alone in a small room with simple, inexpensive equipment, can assemble a film of any desired length from hundreds of separate sections. However, given the same original material, it is unlikely that two film editors would turn out the same type of program.

So long as a single camera is used to expose the originals in short takes from fixed camera positions, these methods of film editing serve the purpose fairly well, and probably it would be difficult to improve on them to any significant extent. They are not well suited to television program production, mainly because it is difficult for the producer to exercise really effective control over the assembly of his program.

Planning Film Programs

Too often program production on film is not planned with the same degree of care and attention to detail as television production. As a rule, television studios and video tape facilities must be booked in advance, and the amount of time that can be allowed for a program is strictly limited. In contrast, film production often takes place outdoors, with a very moderate amount of equipment and a small crew. Even in the larger, more elaborate film productions, one of the main concerns is to give the film editor sufficient footage to allow for any eventuality that may arise during the assembly stage.

This is an extremely wasteful way of working, ill-suited to television, where programs must be made with limited budgets. For every program, even the simplest, a careful plan should be prepared, giving a clear indication in advance of the form the program will take as it is being assembled. With such a plan, it is a relatively simple matter for the camera crew to provide the editor with all of the footage necessary.

When a program is being planned, requirements for titles, special effects such as fades, mixes, and so on, should be included, as the preparation of effects is a slow and costly type of work, especially with color film. A simple fade usually requires an additional duplicating stage in the laboratory, and for the supering of titles an optical printer with register pins is needed to avoid unsteadiness.

A & B Roll Printing Techniques

This method of printing was devised to simplify and reduce the costs of preparing motion pictures on 16 mm film. Simple optical effects, such as fades and mixes, can be introduced into prints by fitting the printer with a device known as a fader shutter.

To make prints by the A & B roll technique, the camera originals are assembled in two separate rolls as shown in Fig. 68. Alternate scenes are interspliced with black leader, so that the A roll, for instance, contains scenes 1, 3, 5, etc, while the even numbered scenes, 2, 4, 6 and so on, are assembled into the B roll.

When a fade is called for, a notch or mark in the edge of the original, ahead of the place where the effect starts, actuates the fader shutter, causing the exposure of the print film to be reduced gradually to zero. If a fade-out and fade-in is to occur between Scenes 3 and 4, the fader shutter would close while the A roll is being printed, at a point near the end of Scene 3. Then, when the B roll is being printed, the fader shutter is programmed to open by a mark placed on the film in the black leader section preceding Scene 4. For a mix, sufficient extra footage must be allowed at the end of Scene 3 and the beginning of Scene 4 to provide the necessary overlap.

The black leader spliced in between scenes prevents light reaching the print film in these sections when the A roll is being printed. Subsequently, scenes in the B roll are printed into these unexposed sections, while the

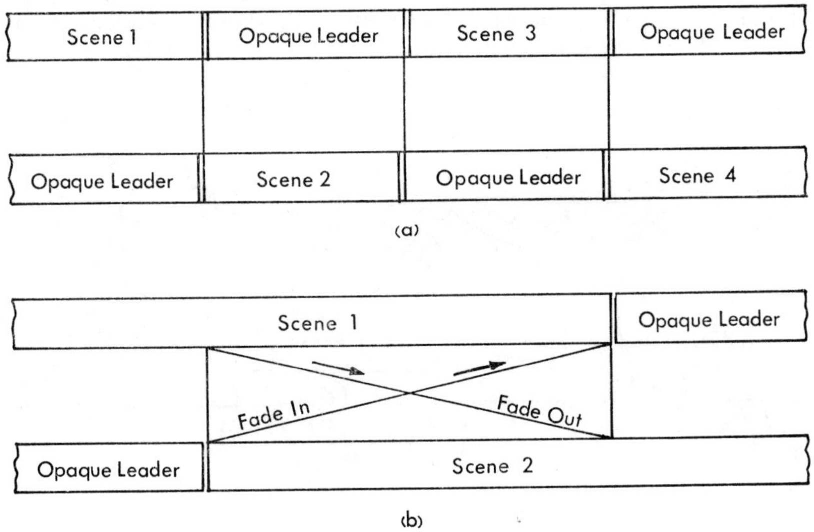

Fig. 68 Preparation of A & B rolls for printing (a) scene-to-scene cuts (b) fade-out, fade-in or mix.

lengths of black leader in that roll prevent light reaching the already exposed portions of the print film.

In addition to simple effects such as fades and mixes, A & B roll printing is frequently employed to hide the unsightly edges of splices in the camera originals. This technique is known as checkerboard printing.

The A & B roll printing method is very effective when only a small number of prints must be made. Although considerable time is needed to prepare the originals for printing, this is the only practical way to obtain 16 mm effects. It has also been used to avoid the costs of optical effects in prints from 35 mm color negatives.

Assembling Multicamera Originals

In a multicamera operation, up to three cameras may be used, running continuously, or started and stopped as required from a central control point. The electronic viewfinders on the cameras give different views of the scene on the picture monitors in the control room. When recording starts the producer selects the camera showing the desired view or angle, and the output of the viewfinder on that camera is switched to a fourth picture monitor on which the successive camera selections are shown in much the same way as the output monitor in a television studio. Simultaneously with the switching of the video signal to the output monitor, a cue mark consisting of a few cycles of audio tone is recorded on the edge of the film in that camera.

The three camera originals, after being processed, are placed on rewinds on an editing bench and threaded into a device known as a synchronizer. This consists of several large diameter sprockets rigidly mounted side-by-side on a shaft. Attached to the shaft is a counter, showing footage and frame

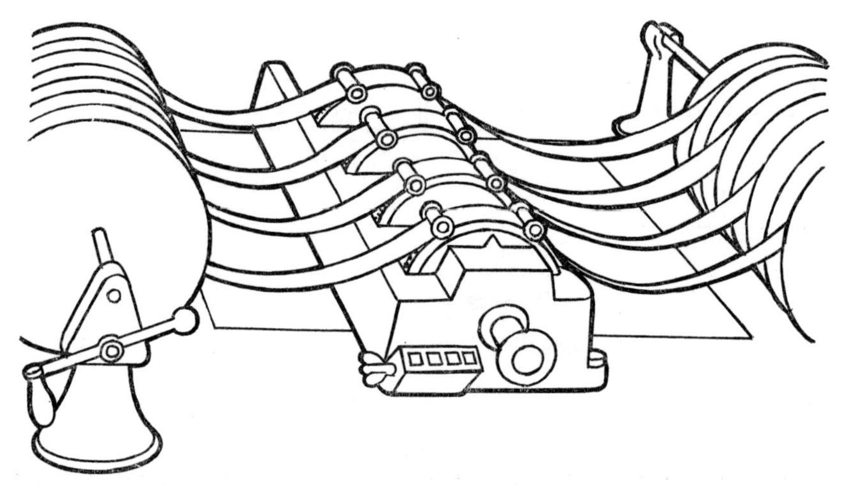

Fig. 69 Assembling a multicamera film program on a four-way synchronizer.

counts, as the film is wound backwards or forwards. A four-way synchronizer accommodates three picture originals and the accompanying magnetic sound film.

If all three cameras are allowed to run continuously for the duration of a take, or a complete program, the three picture films have the same length, and the cue marks recorded on the edges of the films show the camera selected for each portion of the action, as well as the points at which the originals must be cut and spliced to make up the program.

Program assembly can be accomplished very quickly and accurately when the camera originals have been recorded in this way. Editing decisions are made by the producer while the program is being recorded, in much the same way as a television program is produced.

To conserve film stock, the cameras may be started and stopped during the recording operation, with only the selected camera running at any given time. The resulting camera originals will not be the same length, but the cue marks on the originals show the camera that was selected, and these sections may be cut and spliced together using the cue marks as a guide. In making the cuts care must be taken to maintain synchronism between the picture action and the sound on the magnetic film.

Work Prints on Video Tape

A work print is a rough assemblage of print footage taken from the camera originals to serve as a guide for the final cutting and splicing of the camera originals. This procedure safeguards the originals, and at the same time the work print can be taken apart and reassembled as often as necessary during editing operations.

The usual practice is to make rough prints called 'rushes' from the camera originals immediately after processing. The prints are broken down into scenes and takes, each of which is identified at the beginning by a clapper board. The short lengths of film are then hung on pegs in a cloth lined bin by the side of the editor. As each section of film is run through a viewing device, portions are selected and spliced together to make up the work print. The progress of the work can be evaluated by running the completed portions of the work print through the viewer, but it is only afterwards, when the work print is projected on a large screen in the review room that the full impact of the film editor's efforts can be properly appreciated.

The method has also been used for the assembly of television programs on film. The availability of small, low-cost, helical scan video tape recorders should enable film editing procedures to be speeded up considerably. At the same time the program producer can be given the opportunity to participate in the assembly of his program.

Instead of making prints from the camera originals for assembly of a work print, the originals may be transferred to video tape. Using two video tape machines and a simple electronic editor, an edited video tape recording may be prepared. One video tape machine plays back the film transfers, while the other machine records the output of the playback machine. With the aid of an electronic editor scenes may be assembled in any desired manner. As soon as an edit has been made, the recording may be played back immediately to evaluate the effect.

Helical scan video tape recorders are now available with color capability.

Footage numbers may be transferred to the assembly tape while editing is proceeding, to serve as a guide for subsequent cutting and splicing of the camera originals. The numbers may be picked up from a counter dial with a small vidicon camera and inserted into the picture signals while the film transfers are made.

In attempts to simplify and speed up video tape editing operations, electronic engineers have developed equipment for recording time codes in the video tape cue track. These coded signals represent real or elapsed time, and are being used to start and stop machines, and make electronic edits automatically at pre-determined points in the recordings. These coded signals could be recorded when films are being transferred to tape, to facilitate subsequent assembly of video tape work prints.

Electronic Editing of Film Programs

Quite often there is a need for only one print from film used for television program production. In the preparation of the print, it may be necessary to make several trials to obtain acceptable color quality for on-air release. After telecast, the print is of little use, and it is seldom worth while to provide storage space for it, as well as for the original materials.

These problems can be avoided by transferring the camera originals to video tape on broadcast-quality quadruplex recorders, and electronically editing the video tape transfers to make up the final program ready for transmission. As the transfers are made from telecine to video tape, color balance and video levels may be adjusted if necessary. If the first transfer is not satisfactory, the work may be repeated at no cost for materials—the video tape is simply erased, and a new recording is made on it.

All of the camera originals, or selected takes only, may be transferred to tape. Preparation of the originals for transfer requires only rough splicing. When the camera originals are color negatives, it may be desirable to make color prints first, although positive picture transfers could be made from color negatives by polarity reversal of the video signals.

With best possible color picture transfers on video tape, assembly of the program can begin, using the highly sophisticated electronic editing equipment developed for television program production. This equipment includes facilities for previewing transitions between scenes before the actual edits are made. And of course, edits can be remade easily, if the producer so desires, by going back to the original recordings.

Fades, mixes and a variety of other optical effects can be introduced electronically while the program is being assembled.

Assembling Film Programs in Telecine

Program assembly by telecine methods is specially suited for use with camera originals from a multicamera operation. To assemble sets of three camera originals, three color telecines are needed, one for each roll of original picture film. A separate magnetic film recorder in the playback mode reproduces the program sound. A broadcast-quality video tape machine is also required to record the combined telecine output. All of these machines must be electrically interlocked, to run from start marks in exact lip synchronization. Facilities must be provided also to permit all machines to be started and stopped from a single switch in the control room where the program is being assembled.

A basic requirement is a group of precisely calibrated color telecines, set up in advance to give as nearly as possible identical color pictures from test films. The procedures that have been devised to achieve this condition are described in Chap. 14.

If the cameras used to expose the originals were allowed to run continuously throughout the recording of the program, preparation for telecine assembly is quite simple. The picture start marks placed on the films at the beginning of the original recording are located in the picture gates of the telecine projectors to achieve frame-to-frame synchronization. On the other hand, if the cameras were started and stopped during recording, the rolls of camera originals must be filled out with black leader to the same lengths, maintaining lip synchronization with the accompanying magnetic sound track.

A test object such as a gray scale recorded at the head end of each camera original enables the telecine operator to make any adjustments that may be necessary in color trim controls and video levels, relative to the test materials used in the calibration of the telecines. With properly controlled film exposure conditions at the time of the recording, it should not be necessary to make any further adjustments in telecine camera controls during the transfer session.

The video and audio outputs from the telecine machines are fed into a control studio provided with the usual television monitoring facilities. When the start switch is depressed, pictures from each of the telecines appear on the control room monitors, and the producer, working from a script, cuts back and forth from one telecine output to another, in exactly the same way as if the pictures were coming directly from television cameras. First, there should be a rehearsal run, during which the producer may refine the timing of cuts and effects between scenes. Then the final transfer is made to video tape. Afterwards, the tape may be played back to check the assembly.

Titles can be added, and if necessary, commercials can be inserted into the program using another telecine or video tape machine while the transfer is being made.

The amount of equipment required to assemble a program in this manner is quite impressive and, of course, the cost of the equipment is far greater than even the most elaborate film editing facilities. However, this method of program assembly can be justified easily by a much higher rate of productivity, and savings in film materials and laboratory work.

13 Evaluating color television films

In the motion picture industry, it has been the practice to screen all films—including those intended for television use—in review rooms designed originally for evaluating theatre films. There are two very good reasons for following this practice. In the first place, film makers are familiar with the viewing conditions in their review rooms, and good quality color pictures projected on a large screen in a darkened room produce highly favorable visual impressions. Perhaps even more important, film makers are appalled by the lack of uniformity of picture viewing conditions in television stations—differences in the setting of picture monitors and the variability of film reproduction in telecine.

Television engineers have been discouraged in their attempts to improve film reproducing conditions by the extreme variability of film supplied for broadcasting, and by the apparent inability of the film industry to agree on a 'normal' or average color film, that could be used in setting up telecines. Designers of telecine cameras have taken as their objective the most pleasing reproduction of available color films. This has led to the adoption of different color analysis characteristics in color cameras from different manufacturers—a situation in which uniform film reproduction cannot be expected.

Film Viewing Conditions

The normal practice in 35 mm professional motion picture theatres is to project the color pictures with carbon arcs, or a xenon arc emitting light with a color similar to that of the carbon arc. The color temperature rating of the carbon arc is approximately 5400 K. This is close to average daylight in color.

Laboratories making 35 mm color prints provide review rooms simulating theater projection conditions, to enable color prints to be evaluated in a viewing condition similar to a theater. Review room projectors are fitted with light sources with color temperature in the 5400 K range. The luminance at the center of the screen with no film in the projector gate is normally set at 14 to 18 ft-lamberts (48 to 62 nits) to provide a satisfactory viewing condition for color pictures seen in a darkened room. It is recommended that viewers should be located within two to four picture heights from the screen. For a review room seating 10 to 20 people, a normal requirement, the screen size would have to be quite large. A picture height of 8 ft is common.

Because there is not the same degree of uniformity in 16 mm projection conditions, it is more difficult to establish representative viewing conditions in laboratory review rooms for this film size. A luminance level similar to that for 35 mm is recommended for 16 mm as well—14 to 16 ft-lamberts. However, until recently, most 16 mm projectors have been fitted with tungsten lamps, rated at a color temperature of approximately 3200 K. (The color temperature of the source in these projectors is always modified to a considerable extent by the optical system—mainly the heat-absorbing filter, which imparts a blue-green cast to the light.)

Recent industry recommendations* recognize two main groups of 16 mm projectors—those fitted with tungsten lamps in the color temperature range of 3000 to 4500 K, and another group in the 5000 to 6500 K range. Xenon arcs and other high color temperature sources are used to an increasing extent in these projectors, mainly to give higher screen luminance than the old projectors.

Print Color Balance

When a color print is being made, the color balance is adjusted to produce acceptable picture appearance in the laboratory review room. Long experience in making 35 mm color prints for theaters has established an easily-recognizable print balance for this class of work.

In 16 mm film production, the usual practice is to view the prints with a tungsten light projector, and the color balance in printing is adjusted for this viewing condition. However, the demands in 16 mm motion picture exhibition are far less onerous than 35 mm theater requirements, and there is a tendency for less care to be taken in this class of work. Besides, it is much more difficult to maintain color balance in 16 mm prints owing to the all-too-common lack of proper lighting and exposure control of the camera originals when 16 mm stock is used at this stage. When the 16 mm prints are obtained by reduction printing from 35 mm camera originals, at least two intermediate color duplicating stages become necessary, with all the attendant risks of color degradation and distortion.

* USA Standard PH22.100—1967, *Screen Luminance and Viewing Conditions for 16 mm Review Rooms.*

10 2opp.

It is generally believed that a 16 mm print balanced for tungsten illumination should have a distinctly bluish appearance to compensate for the yellow projector light. Some motion picture laboratory workers insist that projector light sources with a color temperature of 5400 K are too blue to give best possible color picture appearance, and 35 mm prints should have a yellowish balance to compensate for this condition. Much depends, of course, on the characteristics of the color images in the film. A 16 mm print compared with the same scene on 35 mm is likely to have a bluish appearance when examined over an illuminated surface, but again, the color temperature of the viewing illumination has to be taken into account. It is known that some types of scene give a pleasant appearance when projected on a screen with either tungsten or arc light. No doubt eye adaptation is more complete when film images have a neutral gray scale, especially in the picture shadow areas.

Effects of Print Color Balance in Telecine

A common practice in setting up a color telecine chain is to adjust the individual channel signal amplitudes to the same level on the waveform monitor, with open projector gate. Alternatively, a neutral gray scale slide or film may be used, the channel gains being adjusted to obtain staircase displays that can be exactly superimposed.

When a color film is then placed in the gate of one of the projectors, signals are produced and displayed on the waveform monitor indicating the amounts of red, green and blue in the color film images. If the images are not neutral the amplitude of one of the waveforms is less or more than the others, and of course, the pictures on the color picture monitor have a corresponding color cast.

A telecine chain with a 35 mm projector on one side and a 16 mm projector on the other—a familiar arrangement in some of the larger network centers—gives waveforms with different amplitudes, and color pictures that are different in appearance, as a result of the methods employed in the laboratory for making these two prints. Some improvement can be achieved by placing a compensating filter in one of the projector light paths. However, this is likely to affect picture colors adversely.*

Film Evaluation Practices

The usual practice in making television programs on film is to evaluate the films in review rooms intended for viewing motion pictures. Film makers are more familiar with their own viewing conditions, and in any event, to purchase, set up and maintain television equipment for viewing films, so as to simulate the use for which the film is intended, would strain

* A recommendation of the European Broadcasting Union, Tech. 3087-E, 1968, specifies that all films for television, 16 as well as 35 mm, will be viewed with a light source in the color temperature range 5400 K ± 400.

the resources—financial and technical—of all but the largest laboratories. There is an even more compelling reason for this practice: through long experience in evaluating films in laboratory review rooms, film makers have acquired a considerable degree of skill in relating the visual appearance of projected pictures to the nature and extent of color corrections required in printing.

The fact that many color films give poor—or at least less than satisfactory—picture reproduction in telecine has been well known for a long time. This problem has become more critical with the advent of large-scale color broadcasting. Many film makers are convinced that there must be something wrong basically with the television system, when films that are acceptable in the laboratory review room fail to give acceptable color pictures when viewed on a television receiver. On the other hand, television broadcasters generally have been quite sceptical of the ability of the film industry to turn out programs with acceptable standards of color quality. Especially in 16 mm telecasting, films show excessive variations in color balance scene-to-scene and film-to-film. Other common faults are excessive contrast, noisy pictures, compression of the picture tonal scale especially in shadow areas, and faces that are too dark or too light relative to picture whites.

What are the reasons for this unsatisfactory situation? Surely the television system cannot be blamed for creating faults that obviously already exist in the films. The television viewing situation is more critical, because the color pictures are seen in lighted living rooms, in comparison with other colored objects. Besides, in making up television programs, films from many different sources have to be spliced together, and any differences between these films, especially in color balance, become glaringly apparent to the viewer.

Viewing Films on Telecine

It would seem that the ideal way to evaluate films intended for television is to reproduce the films on telecine and view the pictures on a television receiver. Many film makers have approached their local television stations seeking assistance of this kind, but most have come away from these sessions frustrated and confused by the advice given them by the television technicians. In many cases, video levels, color trim controls, and sometimes picture monitor controls as well, are adjusted while the film is running to modify the appearance of the color pictures, presumably in a more favorable direction. Seldom is any attempt made to provide reproducible film viewing conditions which would enable film makers to properly evaluate changes that they may have been making in the prints.

There is another quite serious obstacle confronting film makers searching for ways to produce better television films: which of the different types of telecine in use should be used for the evaluation? The color analysis characteristics of telecines from different manufacturers are by no means the same. For example, it is not likely that a flying spot scanner will reproduce colors in the same way as vidicon telecine. To complicate this situation still further, telecines from the same manufacturers show significant differences when measurements of their color analysis characteristics are made.

Review Room for Television Film

Eventually, simple inexpensive equipment may be developed to enable color films intended for television to be evaluated electronically. For the present, an alternative method is to set up viewing conditions simulating television reproduction of the film. These conditions are quite simple to

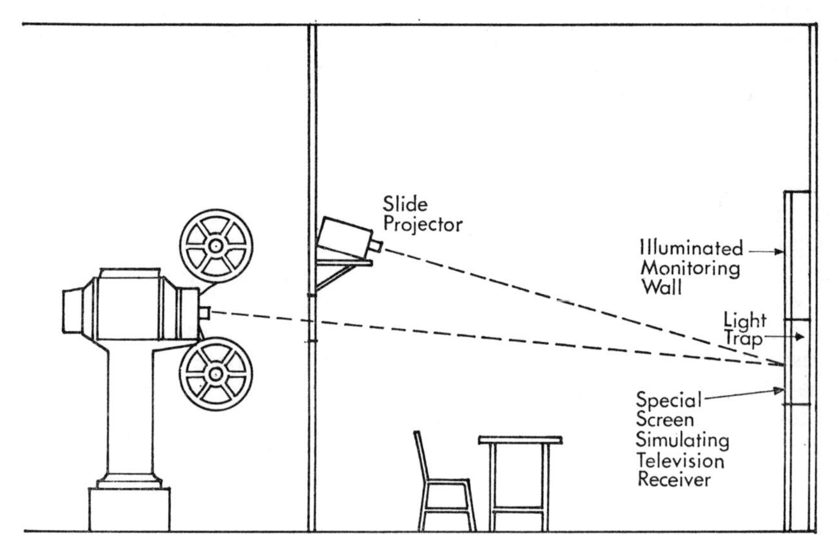

Fig. 70 Review room for television films. The slide projector is used to provide surround illumination.

achieve, and much less expensive to construct than a conventional motion picture review room.

A basic requirement in a television film review room is a projection condition capable of providing a luminance level of 40 ft-lamberts on a screen about 25 in. wide. The reflectance of the screen should be about 20 per cent, comparable with the reflectance of the face plate of a color picture tube. Ideally, the light reflected from the screen should have a color temperature of approximately 6500 K—the generally accepted figure for color television control room monitors—but screen illumination in the range 5000 to 6500 K is considered to be acceptable.

Assuming that films to be evaluated will have a minimum density of about 0.30, the recommended screen luminance of 40 ft-lamberts with open projector gate would be reduced by about one-half, to 20 ft-lamberts in the lightest picture areas with film running. This is about the same luminance level as a properly adjusted picture monitor in a television control room.

The screen should have a large illuminated surround, at a level of about 3 ft-lamberts. The surround light should have the same color temperature as the light from the projector lamp.

When color films are viewed in these conditions, the appearance of the pictures should be similar to the reproduction on a properly adjusted telecine, as described in the following chapter. The illuminated surround prevents eye adaptation taking place. As the level of the surround illumination is raised, the ability of the eye to detect slight changes in color balance increases, but viewing conditions become uncomfortable above a level of about 5 ft-lamberts.

Because the eye is relatively insensitive to fairly large changes in the brightness of picture highlights, it is difficult with this method of viewing to estimate the video signal levels that might be obtained from films. The eye responds easily to a much greater brightness range than the television system can accommodate. Thus in direct projection the eye may see shadow detail that will not be reproduced in the television pictures. However, as skill is acquired in viewing films in this manner, it may become possible to estimate unfavorable film characteristics such as these with a fair degree of accuracy.

A recommended practice of the Canadian Telecasting Practices Committee CTP–1, *Viewing Room for Evaluation of 16 mm Color Films for Television*, describes in detail the requirements for television film viewing, and gives construction details for review rooms. This recommended practice was published in the June 1969 issue of *SMPTE Journal*, page 483.

14 Setting up a standard color telecine

A standard color telecine might be defined as a film reproducing system which can be specified in terms of the video waveforms (and audio signals) obtained from suitable test films and/or slides. Another way to describe a standard telecine is to say that with film reproducing conditions set up in a particular manner, it should be possible to obtain good quality television pictures from film properly made for television reproduction. One of the main advantages of standardized film reproducing conditions is that a film found to be suitable in one telecine will be reproduced in a similar manner in any other telecine. This would provide film makers with a well-defined objective to aim for in producing films for television.

In the past there have been a number of fairly strong arguments against standardization of reproducing conditions for television film. One—perhaps the most convincing—is the existence of a great many dye systems in color films, each with its own peculiar characteristics. According to this argument it would not be realistic to expect that a color telecine could be designed capable of reproducing all of these dye systems without serious errors and distortions.

Another argument is that there are several different telecine systems in existence, each with its own peculiar color analysis characteristics, and it would not be reasonable to expect that all telecines would be modified to conform with the standard, simply to ensure the same film reproducing conditions in all television stations.

There is another, and perhaps even stronger, argument against telecine standardization—why adopt standardized film reproducing conditions when the film industry appears to be incapable of producing films with uniform characteristics?

Problem of Film Variability

It is quite true that color films vary over a wide range in color balance, maximum and minimum densities, contrast, and other characteristics. These variations are due mainly to the use of subjective methods in the preparation of programs on film. The practice of evaluating films in a darkened review room undoubtedly accentuates the variability problem. In the first place, the eye tends to adapt to the balance of the projected pictures. Moreover, individuals viewing films do not all have the same taste for color, or degree of ability in judging color balance. It is very seldom that all of the elements making up a color television film program, including the commercials, can be previewed and remade to achieve uniformity of color balance throughout.

There is the problem, too, that many films used in television programming were made originally for showing by direct projection. Much of this type of film material is supplied to television stations at fairly low cost, and as a rule the quality is quite poor.

Adopting a standard telecine should help to alleviate these problems. The usual practice in the past has been to try to compensate for color variations in the film by adjustments of telecine camera controls. With a standardized film reproducing condition as a reference, the nature and extent of the variations in films would be logged as faults, and steps could then be taken to have these faults corrected. This always has been the practice in broadcasting other types of program materials.

Problem of Different Telecine Systems

In the past, designers of color telecine cameras selected color analysis characteristics (dichroic mirrors and filters) that appeared to give the best television pictures from films known to have good color quality. These subjective judgments were made usually by relatively small groups of individuals, based on the evaluation of a limited number of films. Some vidicon telecines were fitted with wide-band red, green and blue filters to minimize the effects of different film dye systems in the reproduction of colors. Others employed narrow band filters to increase color saturation and reduce dye 'cross-modulation'. In the case of flying-spot scanners, the choice of spectral bands to perform the analysis is limited by the availability of phosphors with suitable spectral energy distributions.

With different types of telecine in use—sometimes two or three types in the same television station—it is simply not possible to reproduce a particular film in the same way on every station; in fact, it may not be possible to obtain similar color pictures from the same film telecast by a particular station on several occasions. So long as variability in films was considered to be unavoidable, there was little concern for these problems—differences in the appearance of color pictures during telecast could be attributed to the film. Now, however, there is a growing realization that any attempt to improve the uniformity of film programs is rendered ineffective to a large degree by the obvious lack of uniformity in reproducing conditions. Film makers can argue: why spend time and money remaking films when television stations are unable to recognize or appreciate improvements in color uniformity?

Once agreement has been reached on the color analysis characteristics of a standard telecine, it is likely that the television industry will quickly adopt the standard, and modification of existing telecines can be started. Reconciling the differences between flying-spot scanners and tube-type telecines may be quite difficult, but no doubt a satisfactory solution for this problem will be found as well.

Problem of Different Film Dye Systems

When a color film is used as an intermediate stage between the original scene and the color television reproducing system, color analysis must be performed twice—first in the film, and then in the telecine camera. In the first analysis the scene is divided into three black-and-white images by the color sensitivity characteristics of the three layers in the film. These layers have relatively broad sensitivity bands, as shown in Fig. 33. Next, dye images are formed in the three layers to produce the color pictures in the film.

It is worth emphasizing at this point that color film materials are not designed to yield accurate color reproductions of original objects. Extensive research has indicated that the eye will not accept accurate reproduction of some colors, especially skin tones. The aim in making color pictures with film is to produce pleasing representations of colored objects and scenes.

For color film reproduction in telecine, the color pictures are again analyzed (separated) into three video signals representing all the red, green and blue light transmitted by the film. As agreement has not yet been reached on the analysis characteristics required in a telecine camera, filter pass bands are selected empirically by telecine designers, to obtain pictures with the most pleasing appearance from good quality films. The problem of filter pass-band selection is complicated by the necessity for reproducing any existing color film. A considerable number of different dye systems have been used in color films. The spectral transmission characteristics of Technicolor dyes, for example, are not the same as the dyes generated in a three-layer film containing color couplers.

There is a rather interesting aspect to this argument. In viewing color pictures the eye does not distinguish differences in the dye systems from which the color pictures are formed. This can be demonstrated easily by taking pictures of a familiar scene with several types of color film. When these pictures are laid down side by side over an illuminated surface, some differences in the appearance of the colors in the pictures may be observed, or the pictures may have slightly different overall color casts, but it is quite likely all of the pictures will be acceptable to the eye as pleasing representations of the original scene. By designing a color telecine to simulate the color analysis characteristics of the normal human eye, it should be possible to reproduce any film dye system in an acceptable manner.

Designing a Standard Telecine

Color films are manufactured, exposed, processed and quality-controlled to give best possible picture appearance in direct optical projection. It could therefore be said that, ideally, the telecine should recreate these pictures on television receivers in viewers' homes with an appearance as nearly as

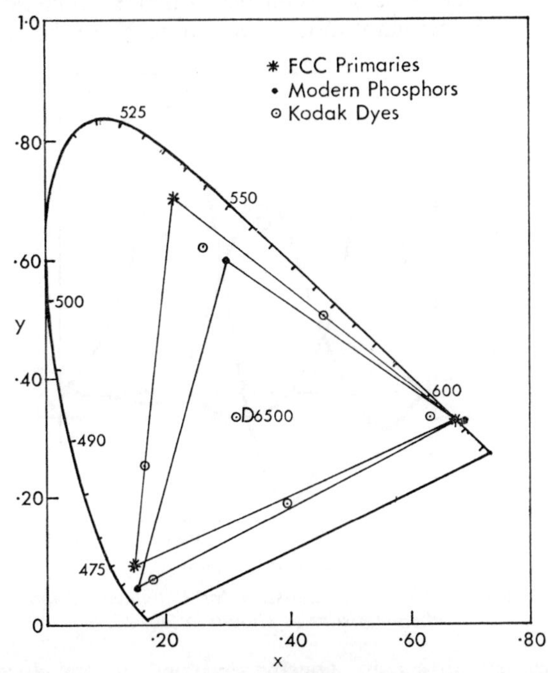

Fig. 71 CIE chromaticity diagram showing the coordinates of typical television receiver tube phosphors and the six saturated colors obtained with the dyes of a typical color print film.

possible identical to the direct projection. Investigations recently have shown that, not only is accurate reproduction in telecine possible, but it is also obtainable in practice. This is a new concept that should prove to be attractive for broadcasters and film makers alike.

The practicability of this concept can be demonstrated by setting up comparison viewing conditions consisting of optical review room projection, as outlined in Chap. 13, side by side with a properly adjusted television picture monitor. Two identical copies of the SMPTE color reference film (or equivalent) are needed to make the demonstration. The telecine used to reproduce one of the copies of the reference film must have color analysis characteristics calculated to reproduce the film accurately. The color analysis characteristics of the telecine must also include the spectral energy distribution characteristics of the phosphors in the picture tube on which the demonstration pictures are viewed.

The optical projection conditions for the other copy of the reference film are designed to simulate the television viewing condition. The color temperature of the screen illumination is approximately 6500 K, the reflectance of the screen is about the same as the reflectance of the picture monitor, and the brightness of the two pictures is similar.

The CIE coordinate system (see Chap. 1) permits a color television system to be engineered in the sense that color performance can be calculated from measurements of the system without having to resort to the judgments of a

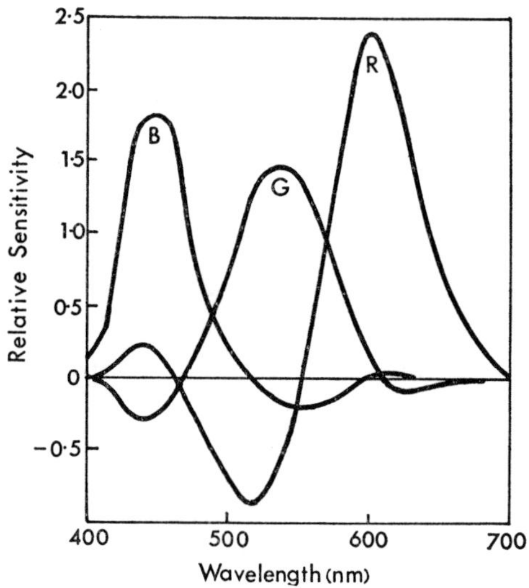

Fig. 72 Ideal telecine taking characteristics for Illuminant D6500 and phosphor characteristics as shown in Fig. 71.

panel of observers. In a color telecine designed to reproduce color films exactly as they appear in direct projection on a screen, the CIE coordinates for each color on the picture monitor must be the same as the coordinates of those colors as seen on the optical projection screen.

The color coordinates of the projected film are dependent on the projector light source. White in color television has been standardized to have the coordinates of Illuminant C* (approx. 6770 K). For these areas to have the same appearance in direct projection as the television picture monitor, the optical projection system must have the same coordinates.

The chromaticity coordinates for the phosphors used in modern television picture tubes are shown in Fig. 71, plotted on a CIE chromaticity diagram. Also shown on this diagram are the color coordinates of the dyes of a typical color print film.

* Illuminant C has been superseded by the more recent CIE designation, Illuminant D, with a color temperature of 6500 K.

Calculating Ideal Telecine Color Analysis Characteristics*

Knowing the spectral energy distributions of the projector light source and the phosphors in color picture monitors, it is possible to calculate the characteristics that the red, green and blue channels in the color telecine camera must have for accurately duplicating the directly projected images. Fig. 72 shows the spectral analysis curves derived by mathematically transforming the plane of the television primaries into the plane of the CIE primaries.

It should be possible to duplicate the positive portions of these curves with available filters, but to simulate the negative portions, light would have

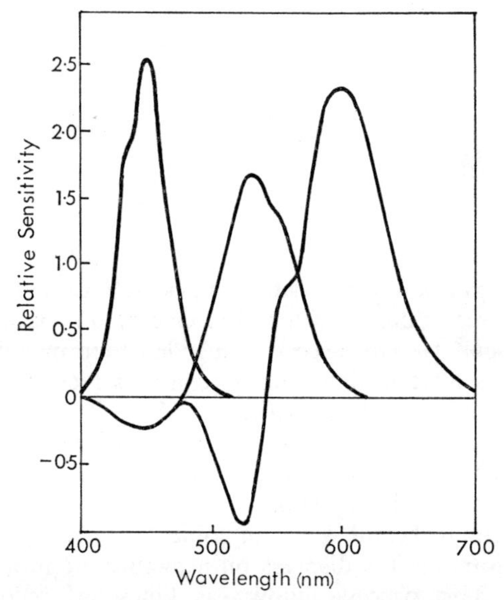

Fig. 73 Practical telecine taking curves obtained by utilizing available trimming filters and electronic matrixing.

to be subtracted. This type of effect can be achieved by electrically subtracting one color signal from another, using a matrixing amplifier.

To determine the best combination of matrix coefficients and camera filters, a test object made up of saturated film dyes may be used, and errors in the reproduction of the colors may be calculated in JND (just noticeable difference) units. Using this procedure, it should be possible to modify any film reproducing system to obtain color television pictures closely resembling the appearance of the directly projected pictures as seen in a standard television film review room.

* Under the general heading *An Engineering Approach to Color Telecine*, three papers by Quinn, McRae and Corley, describing the procedures used to establish standard film reproducing conditions, appeared in the March 1969 issue of *SMPTE Journal.*

15 Transferring color video tape to color film

For several reasons it is highly desirable to be able to make transfers from color video tape recordings to color film. One of the main reasons is that it is always difficult to gain access to a video tape machine for viewing recordings. Copies of recordings on film can be screened much more conveniently on a motion picture projector. Moreover, the distribution of television programs on video tape is hampered by differing television scanning standards around the world. Film, on the other hand, can be reproduced in any television system.

In the earlier monochrome era, film copies of television programs were made by photographing the displays on a picture monitor with a motion picture camera. This process, known as kinescope recording, or telerecording, produced generally low grade results, compared with conventional motion picture films, and usually, when film recordings were compared with the original television pictures, severe losses in picture quality could be readily observed. With the advent of video tape recording in 1956, interest in the kinescope recording process rapidly waned.

Attempts to transfer color video tape recordings to film have been more successful. While color film transfers are much more difficult to make, the picture quality that can be achieved is superior to the best monochrome transfers, in direct comparison with the original pictures from video tape and with conventional color films. The success of the ventures that have been undertaken so far has encouraged the development of new equipment and techniques for making color film transfers, and it is quite likely still further improvements will be made in the future.

The most urgent requirement is improvement in uniformity of the transfer process. The most obvious disadvantage of kinescope recording has always been variability in picture quality, due to the extreme susceptibility of film to slight variations in recording conditions, and the many factors affecting the film recording process. An automated film transfer system in which all of these factors can be precisely controlled, would appear to be the answer to this problem.

Outline of the Film Transfer Process

The simplest way to make a transfer from video tape to film is to set up a motion picture camera facing a color picture monitor, the images appearing

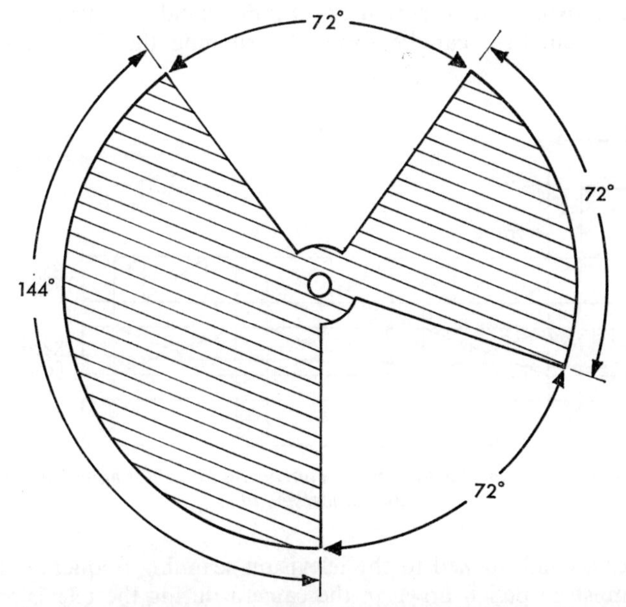

Fig. 74 Special camera shutter for alternate field recording.

on the picture monitor being sharply focused by the lens at the plane of the film in the camera. Any color film suitable for original photography may be used, but high speed reversal materials are favored.

A recording of a television display made with an ordinary motion picture camera shows vertically-travelling horizontal bars, owing to interference between the television scanning rate and the motion picture frame rate. In a film camera, a rotating shutter admits light to the film while it is at rest, during the exposure period; then in the period when the film is being advanced, the opaque sector in the shutter blade cuts off the light. At the standard motion picture frame rate of 24 per second, about half of the total time period for each frame—41.5 milliseconds—is taken up in moving the film, while exposure takes place in the remainder of the period—nominally 1/50 sec. or 20 milliseconds.

In the North American television system the scanning rate is 30 frames per second—about 33 milliseconds per frame—with each frame being made up of two interlaced television fields. Several different methods have been devised to avoid interference between the two frame rates. The simplest method is to use a shutter in the camera with two open sectors as shown in Fig. 74. When this shutter is rotated at 720 rpm one television field is recorded during each 72° shutter opening, producing what is known as an alternate field recording. Only 40 per cent of the picture information is recorded. Equipment for full-frame recording drops out a half-field between film frames, during which film pull-down occurs.

The European CCIR television system has 25 frames—50 fields—per second, occupying a time period of 40 milliseconds per frame. The interference problem here can be solved by running the film camera at 25

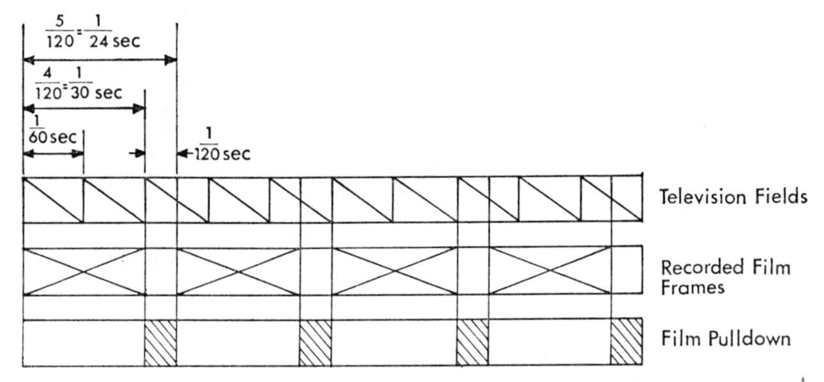

Fig. 75 Timing diagram illustrating the conversion from 30 frames/sec. television to 24 frames/sec. film.

frames per second, locked to the television scanning frequency. However, the film must be pulled down in the camera during the television vertical blanking interval—a period of less than 2 milliseconds, to avoid loss of scanning lines at the top or bottom of the film frames.

With any method in which film pull-down occurs during vertical blanking, the rotation of the camera shutter must be locked to the television synchronizing generator.

Film Exposure Control

The display on a color television picture monitor is made up of variations in the brightness (luminance) of large numbers of tiny phosphor dots. For example, if the blue area of a picture on the monitor is examined, it will be found that only the blue phosphor dots are emitting light, while the red and green dots cannot be seen. Ideally, only the light from the blue dots should produce images in the blue-sensitive layer of the color film, while the red- and green-sensitive layers remain unaffected. Similarly, the red phosphor dots should produce images only in the red-sensitive layer of the film, and the green dots should affect only the green-sensitive layer.

These ideal conditions are very difficult to achieve in practice. If a television signal producing red, green and blue color bars on a picture monitor is recorded on color film, images are formed in all three layers of the film, because the color sensitivity of the individual layers is quite broad, and at the same time, emission of energy by the phosphors is not restricted to narrow bands in the spectrum. The spectral energy distributions of typical phosphors used in typical television picture monitors are shown in Fig. 76, together with the color sensitivity characteristics of a typical color film.

The effects of exposure in the three layers of the film must be taken into account, because the appearance of the film pictures is influenced by these undesirable exposure effects. Exposure may be adjusted to obtain the most pleasing film pictures through the use of filters over the camera lens. Alternatively, the pictures appearing on the color monitor may be altered by appropriate adjustments in monitor set-up, or by modification of the signals applied to the monitor, using matrixing techniques.

The level of exposure should be adjusted to reproduce the color pictures in the central straight-line portions of the film's characteristic curves (in a color film each layer has its own characteristic curve). This can be done by displaying a monochrome staircase signal on the picture monitor, and recording it on the film at different levels of exposure (lens apertures). A suitable photometer is needed to adjust the luminance of the white step of the staircase to an appropriate value, as well as the difference in luminance values between the white and black steps (contrast). In this way, camera original films may be produced with any desired picture contrast characteristic, for a given color processing condition.

Color Processing Control

The key to successful color recording is precise reproducibility of processing conditions. Any variations in processing upset the calculations for exposure of the film. Uniformity of processing can be checked by making sensitometric exposures in the color film, the resulting images being measured with a color densitometer. These measurements, plotted on a process control chart, show the nature and extent of any variations that may occur (see Fig. 44).

Alternative Recording Methods

Among the more promising of other methods that have been used for recording are the following:

1. Area sharing of color separation images on black-and-white film.
2. Lenticulated film systems using black-and-white emulsions on a specially processed base material.

3. Triniscope systems with three picture tubes, the images being multiplexed optically.

4. Separation negative process, in which three separation negatives are made on black-and-white films. To obtain prints, the Technicolor dye transfer method may be used, or the separations may be printed one after the other on color print film.

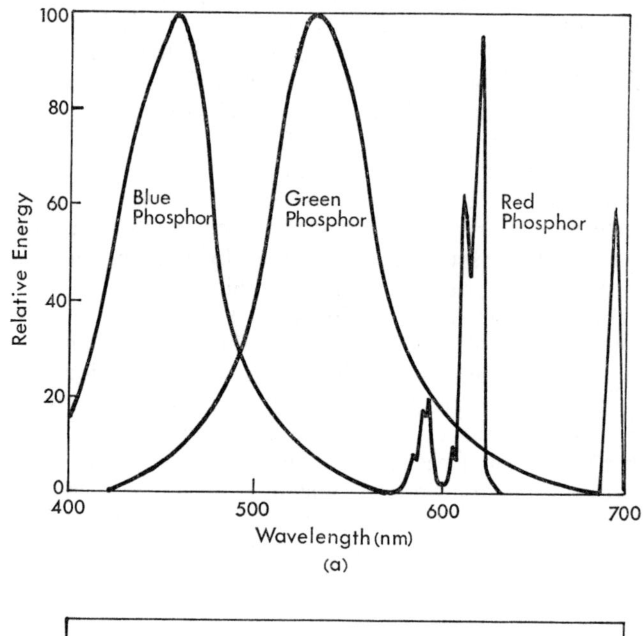

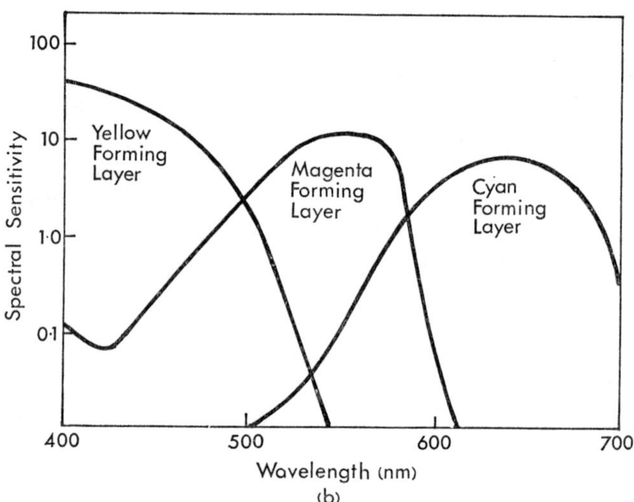

Fig. 76 (a) Spectral energy distributions of typical picture monitor phosphors and (b) spectral sensitivity curves for the three layers in a color camera film.

In the triniscope system the color video signals are decoded to obtain the R, G and B components, which are fed to three picture tubes. The images appearing on the tubes may be modified by special signal processing techniques, and optical filters may be selected to obtain the desired spectral energy distributions in the three colored light beams entering the camera. One of the most difficult problems with systems of this kind is precise registration of the three colored images at the film plane.

The separation negative process was used many years ago in the production of Technicolor motion pictures. Of all the systems that have been devised for making color transfers this has the greatest possibilities. While registration is always a problem in any separation process, the system offers many important advantages—mainly precise control of color image formation and economical production of large numbers of prints.

Recovering Picture Information from Video Tape

In many ways the advent of video tape recording made film recording operations much easier. Instead of recording the live television pictures directly on film, a video tape recording could be made first, and if necessary, transfers from tape to film could be repeated to obtain the best possible results on the film.

While these transfers are being made, the video signals recovered from the tape may be enhanced or modified in any way that appears to improve the quality of the film pictures. Perhaps even more important, the recordings on video tape provide a reference against which transfers to film may be compared.

Subjectively, as viewed on a television picture monitor, pictures recorded on video tape are usually so good that it is difficult to distinguish the recording from the original pictures. When transfers are made to film, however, it soon becomes obvious that video tape recordings contain many faults and defects. As a rule, the film transfer process exaggerates these faults; sometimes the tape faults seriously interfere with the operation of the film recording equipment.

The signal-to-noise ratio of the signals recovered from video tape is a major factor in obtaining high quality film transfers. Noise in the video signals becomes greatly magnified and enhanced during the transfer and contributes greatly to the grainy appearance of the pictures as recorded on film.

The type of lighting employed in the original set or scene is another important factor. Flatly-lit scenes produce film transfers with limited tonal scale, and there is very little that can be done to correct the condition in the transfer process. Ideally, the video tape should yield signals with minimum noise, and the pictures should have sufficient key lighting to enhance subject contrast range. Extreme long shots should be avoided owing to the marginal resolving power capabilities of current transfer processes. Close-ups should be emphasized. Chroma levels should not be excessive, as color saturation tends to increase, especially when reversal film stocks are used for camera originals and prints.

By far the most important factor in setting up color balance and exposure levels is the reproduction of skin tones. Inanimate objects, such as commercial products and scenery may pass as quite acceptable with small shifts from true coloration, but flesh tones with an 'unnatural' appearance cannot be tolerated.

In some cases scenes in a recording may not match the color bar signals recorded at the head end of the video tape. Worse still, there may be serious color mis-matching between scenes or shots, because of faulty camera line-up prior to the original recording, or the editing together of recordings made at different times with different television cameras.

One of the most annoying faults in video tape recordings is 'color banding'. This type of fault appears as horizontal stripes of different colors across the picture, especially in highly saturated areas. All video tape recordings have picture defects caused by imperfections in the tape itself. These defects, called 'drop-outs' may produce white or black horizontal streaks or tearing in the pictures, and in some cases, a more serious disturbance known as 'roll-overs'.

Colorimetry Considerations in Color Film Transfers*

In making a color transfer from video tape to film, exposure conditions must be chosen to give:

1. Best possible reproduction of saturated colors.
2. Neutral reproduction of the picture gray scale.
3. Acceptable picture contrast.

The conditions needed to satisfy these requirements are to some extent interdependent. Color saturation depends on the relation of the density of each dye image in the positive print to the corresponding color signal applied to the recording equipment.

Distortion of the red, green and blue separation signals can occur during recording as the result of the overlapping of the sensitivity bands of the three color film layers. Thus there are some wave-lengths of light to which all three layers will respond to some extent.

To overcome this undesirable condition, the light reaching the film from the color display device must have spectral energy distribution characteristics so arranged that light in these portions of the spectrum is suppressed. Some improvement can be achieved with electronic masking techniques.

A color transfer system employing three separate cathode ray tubes is shown in Fig. 77. With this arrangement, the spectral energy distribution of

* BBC Engineering Monograph No. 72, *Colour Sensitometric Parameters in Colour Film Telerecording* provides additional information on the subject.

the light entering the camera may be controlled by filters placed over the tube faces, and by a dichroic mirror arrangement.

The spectral sensitivities of the three layers of a typical color film used for making film transfers are shown in Fig. 76. It can be seen from this illustration that to expose the red-sensitive layer without affecting the other two layers, the output of the red tube must not contain light with wavelengths shorter than 610 nm. Similarly, the blue-sensitive layer should be exposed with a narrow spectral energy band close to 430 nm for minimum interaction with the other two layers.

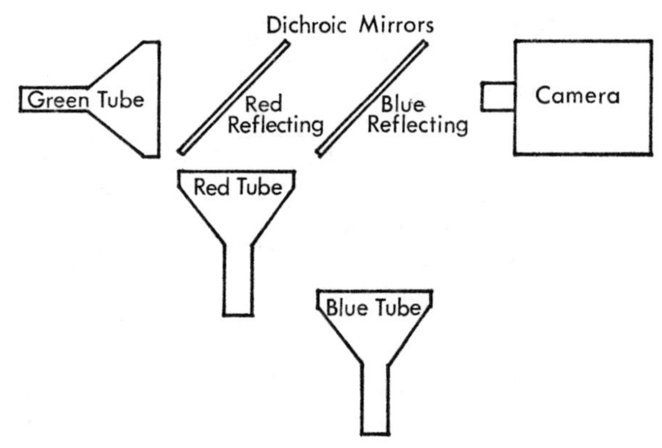

Fig. 77 Film recording equipment using three cathode ray tubes and an arrangement of dichroic mirrors.

It is impossible to expose the green-sensitive layer without affecting both of the other layers to some extent, but wavelengths in the region of 530 nm give the best separation.

After the most favorable tube phosphors, filters and dichroic mirrors have been selected, the efficiency of the recording system can be tested by illuminating each of the tubes in turn, while exposures are made on the selected film stock.

It should be possible to obtain a density difference between wanted and unwanted color images of at least 2.50, or about 50 dB. The separation in the green channel is likely to be somewhat less—perhaps 2.0—but the practical effect of the unwanted exposure is negligible.

Electron Beam Recording

Conventional kinescope recording employs a phosphor faceplate and optical system as an interface between the electron beam carrying the video signal and the film used to record the signal. Electron beam recording brings the film inside the vacuum chamber where the electron beam can be used to expose the film by direct electron bombardment.

This requires a very small spot diameter to achieve maximum horizontal resolution. Assuming equal line widths and no space between lines, the placement of 489 active scanning lines in a 16 mm film frame requires a spot size of less than 15 microns. To reduce the effects of contamination when the electron gun is opened to the atmosphere to load and unload the film, a tungsten filament is used, together with a multi-stage vacuum chamber.

Exposure of the film by an electron beam is almost instantaneous, thus avoiding the problems of phosphor afterglow in conventional recording equipment. Conversion from 30 television frames to 24 film frames is accomplished by discarding one-quarter field in the pull-down period of

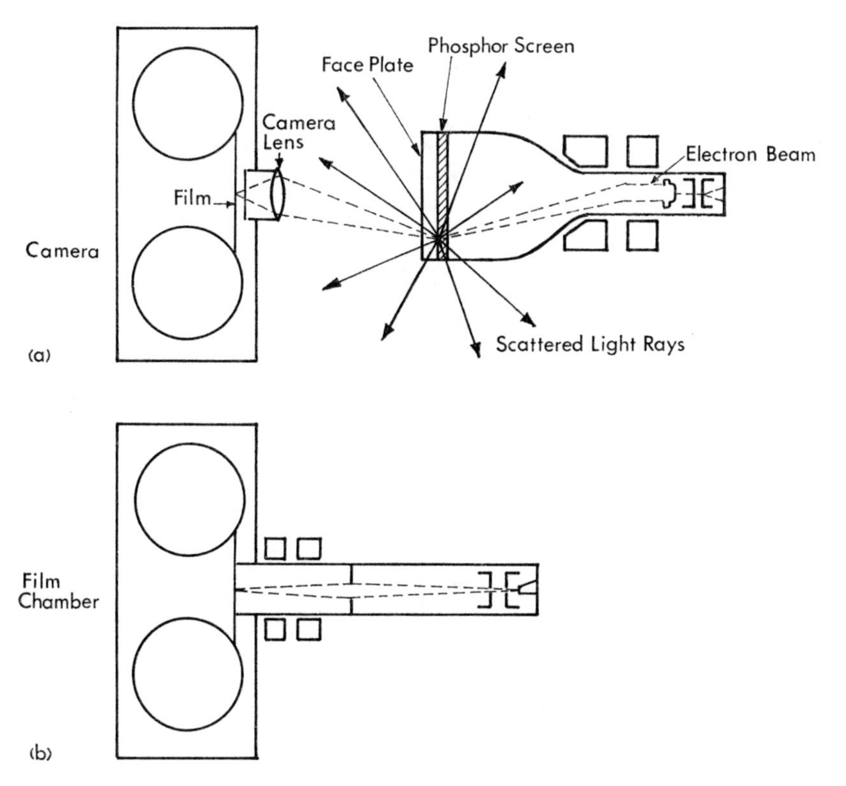

Fig. 78 Comparison of recording systems (a) kinescope system employing a phosphor screen (b) electron beam recording system.

about 8 milli-seconds. Film exposure control is accomplished by metering the electron beam during horizontal blanking when a reference pulse is generated. Electron gun bias is automatically adjusted to maintain the amplitude of the reference pulse. With this method of recording, 16 mm transfers can be made with resolution in excess of 600 lines, free of the fixed noise patterns of phosphor faceplates.

The electron beam recording system has been developed to the practical stage for monochrome recording.* It is likely that a method of color recording using this principle will be made available in the near future.

* For further information on this subject see two papers in the May 1966 issue of *SMPTE Journal*, pp. 191–197, *Television Film Recording using Electron Exposure* and *An Electron Beam Television Recorder*.

Index